DATE DUE

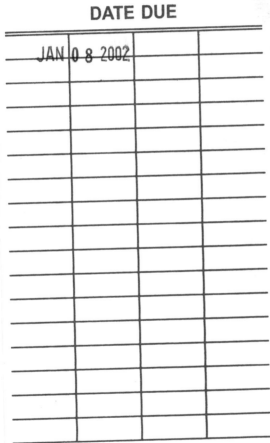

JAN 0 8 2002

Demco, Inc 38-293

Capillary Electrochromatography and Pressurized Flow Capillary Electrochromatography

An Introduction

Ira S. Krull
Northeastern University
Boston, Massachusetts

Robert L. Stevenson
Abacus Group
Lafayette, California

Kavita Mistry
Northeastern University
Boston, Massachusetts

Michael E. Swartz
Waters Corporation
Milford, Massachusetts

HNB Publishing

NEW YORK

ISBN: 0-9664286-2-5

This book is printed on acid-free paper.

The publisher offers discounts on this book when ordered in bulk
quantities.

HNB Publishing
250 West 78th St.
New York, NY 10024
www.hnbpub.com

Current printing (last digit):

10 9 8 7 6 5 4 3 2 1

PRINTED IN THE UNITED STATES OF AMERICA

Preface

Capillary electrochromatography (CEC) has become an area of intense interest, attention, research, and commercial development all around the world. Although CEC has been described in the open literature for almost three decades, its flowering in terms of publications, presentations, reviews, and dedicated meetings is much more recent, perhaps less than one decade.

This book begins by describing the basic operations and principles of CEC, as well as pressurized flow variations, such as pressurized CEC (PEC) and electro-high-performance liquid chromatography (electro-HPLC). We describe the basic instrumentation, principles of operations, basic equations of operation, history, and making of suitable capillaries for conventional, isocratic, or gradient CEC. The problems remaining and opportunities for improving capillary packing technologies are discussed, as well as the proper selection of packings (packed beds vs. monolithic or polymeric). We next present an in-depth discussion of the selection of mobile phases for CEC; how these affect electroosmotic flow; and how the proper buffer, organic solvents, pH, ionic strength, and other buffer components can all affect the total elution time, migration time, dead volume, and overall analysis time.

Chapter 4 emphasizes the instrumentation prevalent today in CEC, what commercial sources exist, how conventional capillary electrophoresis (CE) instrumentation can be adapted for CEC/PEC or electro-HPLC, how buffer and capillary pressurization can be utilized (and why they should be), and the major detectors commonly used in CEC today. The next two chapters discuss different areas of application where CEC has been pursued, optimized, and even applied to real-world samples. Chapter 5 describes applications for small molecules, especially pharmaceuticals, chiral resolution, amino acids, and so forth. Chapter 6 covers applications for biopolymers, including proteins, peptides, nucleosides, carbohydrates, and related compounds.

Chapters 7 covers areas of method transfer from HPLC to CEC or PEC, and the best approaches possible for so doing, with recommendations for use of commercial software. Chapter 8 covers the area of method development and optimization in CEC/PEC, how to approach such an optimization and know when optimization is reached, and the best approaches to optimize a new method in this format.

Finally, Chapter 9 discusses possible future developments in CEC; what is expected to next occur in its evolution, what improvements in capillary technology, detectors, and packings are yet needed; and related areas, including novel applications not yet explored. We conclude with an overview of where this technique has come from, where it is today, and where it might be tomorrow, if it continues to develop in a positive manner and direction. We also provide some warnings for those analysts who might wish to investigate and apply CEC/PEC to their own unique applications areas, and how they might garner more success than otherwise, depending on the particular approaches they take.

The book does not attempt to describe method development based solely on the structures of analytes but rather utilizes adaptation of existing HPLC or CEC conditions and methods for a given analyte or its related compounds. Thus, we do not discuss CEC method development by working strictly from a given analyte's structure, selecting the specific packing and mobile phase conditions, and so forth. We have purposely chosen to start with the literature describing existing CEC or HPLC methods for a particular structure or group of compounds, modify and optimize these conditions as needed for a particular analyte's structure and/or sample matrix, and then proceed with method validation of the final, optimized set of CEC conditions and separations realized.

We are very grateful to a number of individuals who have encouraged us. Through a whirlwind tour of the UK, ISK was privileged to speak with individuals knowledgeable in CEC areas, including Brian Clark, Keith Bartle, John Knox, Mel Euerby, Kevin Altria, Bob Boughtflower, Clair Paterson, Derek Reynolds, and Norm Smith. In addition, various packing materials, technology, techniques, approaches, figures, overheads, preprints, and discussions were provided by Peter Myers of the University of Leeds. Gerard Rozing of Agilent Technologies provided an especially thorough review of the manuscript. His constructive suggestions improved the content of the book.

Within the US, several knowledgeable individuals also provided fruitful discussions, reprints and preprints of publications, and encouragement, including John Stobaugh, Luis Colón, Richard Zare, Chao Yan, Bob Weinberger, John Dorsey, and Vince Remcho. Professor Cs. Horvath of Yale University allowed us to view his laboratory, speak with his graduate students and staff scientists, and learn even more about CEC and his own areas of intense activity and interest over the past several years. Frantisek Svec of the University of California, Berkeley, offered a number of helpful comments about the manuscript.

Waters Corporation provided ISK with a quarter's sabbatical at its facilities in Milford, MA, where he was charged with learning as much as possible about CEC and its possible commercial implications and future. Within Waters, several individuals were very helpful in providing discussions, insights, collaborations, and support, including Bob Pfeifer, John Nelson, Steven Cohen, Michael Swartz, Bob Karol, Devette Russo, Art Caputo, Uwe Neue, Pat McDonald, Ed Bouvier, Jim Krol, and Tad Dourdeville.

Waters also generously provided us with various HPLC packing materials to be used in CEC capillaries and applications, information and literature about peptide mapping in HPLC, and HPCE instrumentation, equipment, software/hardware, supplies, and accessories in support of our CEC R&D efforts.

We acknowledge the support of Merck Research Laboratories (Rahway, NJ), which provided a research grant to Professor Krull's group at Northeastern University, as well as a Merck Career Development Award to Kavita Mistry.

We are very grateful to a number of graduate students, including A. L. Sebag and Sarah Kazmi, who devoted time and effort to researching areas of CEC. Thorsten Lobert worked with ISK for the fall quarter of 1997 at Waters in developing certain areas of CEC. We are most grateful for all of his assistance, contributions, collaborations, and skillful laboratory expertise.

Finally, we are appreciative of the support and encouragement of our publisher. We hope that this patience will be rewarded by the reception that the book receives from you, the audience, who may want to apply the information contained herein for your own unique interests and applications. If there are faults, they are the authors', and not any of the individuals or organizations mentioned above.

Ira S. Krull,
Robert L. Stevenson,
Kavita Mistry
Michael E. Swartz

Contents

1 Introduction

Capillary electrochromatography (CEC) is an analytical separation technique that strives to combine the best features of HPLC and capillary electrophoresis (CE) or high-performance capillary electrophoresis (HPCE). This book is intended to present the principles of operation and to assist the interested scientist with the successful application of the techniques of CEC. The book offers an objective comparison of CEC with two closely related techniques, CE and HPLC, that emphasizes optimizing operating parameters and avoiding misconceptions that could lead to laboratory blunders.

1.1 Some Definitions and Concepts

1.1.1 Electrophoresis

Electrophoresis refers to the differential migration of particles, usually molecules or ions, under the influence of an applied electric field. The rate of migration (u_{ep}) is dependent on many factors, including the applied voltage (E), the dielectric constant of the medium (ε) the zeta potential (ζ), and the viscosity of the medium that the particle is moving in (η) [1]. This is shown in Equation 1.

1

$$u_{ep} = \frac{E\varepsilon\zeta}{6\pi\eta}$$ (Eq. 1)

Because of differences in charge and ionic radius, particles move through the capillary at different rates. Small, highly charged ions move through the capillary faster than, and hence are separated from, larger ions, especially those with lower charge (Figure 1). This difference in migration velocity, u, forms the basis for a CE separation. The output from an electrophoresis run is called an electropherogram. It can be a time graph or a two-dimensional display with spots or bands.

In the mid 1980s, scientists began to recognize that it was possible, perhaps even easy, to improve speed, quantitative analysis, and automation by converting slab gel electrophoresis to a capillary format. The technique came to be called CE or HPCE. With HPCE, it was

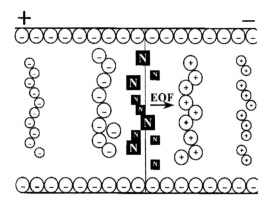

Figure 1. Electrophoresis is the movement of charged species in an electric field. Large molecules with low charge move more slowly than small, highly charged species. This difference in migration velocity, u, is the basis of the separation in the same manner as a foot race. The output from a detector is recorded as a function of time or position. It is called an electropherogram. (From Ref. [2], with permission.)

practical to reduce run times by about 90% due to more efficient heat dissipation afforded by the thin capillary wall. More efficient heat dissipation, in turn, enabled the use of higher field strength (voltage), which shortened the separation time.

1.1.2 Electroosmotic Flow

In a perfect world, electrophoresis would be a simple technique to master; this being the real world, however, there are complications. Electroosmotic flow (u_{eo}) is one complication, which can be exploited. Many surfaces in contact with a conducting liquid have a surface charge. This effect is illustrated in Figure 2.

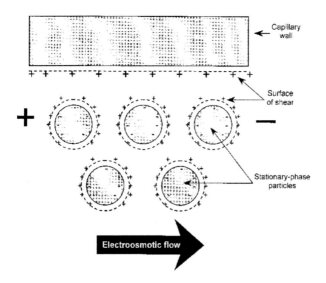

Figure 2. Electroosmotic flow is generated by movement of charges along a liquid–solid interface. Many solids have fixed charges on their surface. In the case of silica, the charged group is $\equiv$ Si-O$^-$. To preserve charge neutrality, there is a cation, such as Na$^+$, associated with the surface charge. This charge is mobile and can be replaced by another cation in solution. In the presence of the electric field, cations are attracted to the opposite charge, called the cathode. As they move toward the cathode, they drag along the water, producing a net liquid flow, which is called electroosmotic flow. (From Ref. [3], with permission.)

Electroosmotic flow is a function of the applied electric field (E), the dielectric constant of the mobile phase (ε), the zeta potential (ζ) (which is the potential difference between the Stern layer and the shear layer of movement [see Figure 10]), and the viscosity (η), as shown in Equation 2.

$$u_{eo} = E\frac{\varepsilon\zeta}{\eta} \qquad \text{(Eq. 2)}$$

The velocity along the capillary can be as great as a few cm per minute and is often greater than the electrophoretic velocity.

1.1.3 HPLC

Chromatography is a separation technique in which the analytes partition differentially between stationary and mobile phases (Figure 3). Compounds that are attracted to the stationary phase are washed down the bed more slowly than those that prefer to stay in the mobile phase. The differential migration, in a so-called countercurrent manner, is the basis of chromatographic separation and theory. The output of a chromatographic separation is usually a time chart recording the detector response as a function of time.

Looking at the details of Figures 1 and 3, the essential difference is that electrophoresis is based on the differences in speed of migration of compounds in an electric field. Typically, only one phase is used. Chromatography is a two-phase process, exploiting the differences in chemical attraction between molecules in mobile and stationary phases. With chromatography, it is possible to change the separation process by changing the surface chemistry of the stationary phase, which can change the strength of the interactions. Stronger interaction with the stationary phase means slower migration along the column. It is also practical and effective to increase the interaction of the sample with the mobile phase, which increases the migration rate along the column.

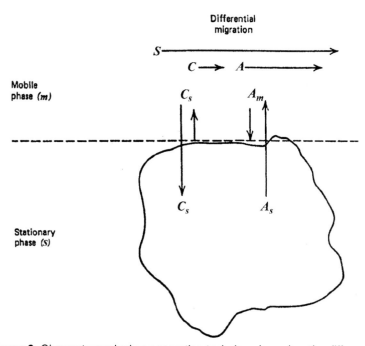

Figure 3. Chromatography is a separation technique based on the differential rate of migration of a sample along a column. The column is usually packed with an adsorbent. The components of the sample are dissolved in the mobile phase, which percolates through the bed. Some compounds (S) in a sample mixture may not interact at all with the stationary phase, and these emerge at a time corresponding to the void volume of the column. Others may experience weak attraction such as (A) and emerge more slowly. Still others (C) may experience strong attraction and elute even later, or perhaps not at all. The recording of this phenomenon is called a chromatogram. (From Ref. [4], with permission.)

While there may be a lack of agreement about precisely what *HPLC* stands for (*h*igh *p*ressure, *h*igh *p*recision, *h*igh *p*erformance, or *h*ighly *p*roductive), the method uses pressure to force the mobile phase through the packed bed of the column, as shown in Figure 4. Resistance to flow can generate very high pressures, in the range of 500 to 6000 psi, or even more for a linear velocity of a few cm/min.

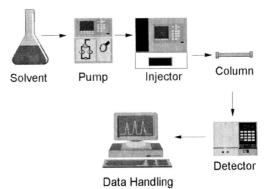

Solvent   Pump   Injector   Column

Data Handling   Detector

Figure 4. Typical HPLC instrument. A solvent management system pumps the mobile phase from a reservoir to the injection valve, where the sample is introduced, and then to the separation column, and ultimately the detector. The output of the detector is recorded as a function of time, producing a chromatogram. (Used with permission of Waters Corporation.)

1.2 Capillary Electrochromatography: The Best of Both Worlds

HPLC is arguably the most commercially successful technology in analytical chemistry. Annual sales are twice that of apparatus for the second largest selling technique (gas chromatography or mass spectrometry). There are about 150,000 active instruments around the globe. Only pH meters and balances are more widely used. So what could be improved?

During the early days of HPCE, many questioned if HPCE might replace HPLC. Rather quickly, it became apparent that HPCE did not have the sample load, concentration sensitivity, or dynamic range of HPLC. The low sample load was particularly bothersome for scientists discovering a new peak. The first question is: What is that? followed by Are your *sure*? Answering these questions with HPCE alone was very difficult.

In the early 1990s, it became apparent that for most aplications HPLC and HPCE were complementary. But then scientists began to realize that packing the capillary with a column packing would improve the amount of sample that could be injected. And if electroosmotic flow were used, the capillary could be much longer than is practical with current HPLC instrumentation. So they found that by substituting electrically driven flow for pressure, it was possible to obtain the efficiency of HPCE with the sample loads of HPLC. The combination of a packed column eluted with flow provided by electroosmosis is called CEC. Some experts will wonder how CEC compares with micellar electrokinetic chromatography (MEKC), which is a well-developed mode in HPCE. The most obvious point is that in CEC one usually uses a stationary phase that is fixed in a bed or on a tube. In MEKC, micelles are formed and move opposite to electroosmotic flow in the buffer.

Many of the same applications that have been developed for HPLC have been investigated by CEC. A typical separation by CEC is shown in Figure 5. The electrochromatogram shows the separation of

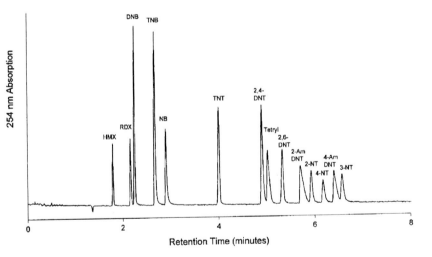

Figure 5. CEC separation of 14 explosive compounds. Column: 75-µm i.d. × 17 cm packed with 1.5-µm nonporous ODS. Mobile phase: 15% CH$_3$OH/85% 10 mM MES (pH 8. 5). Voltage: 12 kV. Injection: 1 kV for 1 sec. UV detection at 254 nm. (Used with permission of C. Yan and Unimicro Technologies, Inc.)

14 explosives in about 7 minutes. The peaks are narrower, and the run time is shorter then expected for the HPLC separation of the same sample.

HPLC is constrained by the number of components in a sample that can be resolved. This is called *peak capacity*. It is a measure of the number of peaks that can be placed in an ideal chromatogram (Figure 6). The practical peak capacity of HPLC is less than 25 peaks. Even under ideal conditions, peaks may arrive at the detector at the same time, which is called *co-elution*. However, since CEC has a peak capacity about 10 times larger than HPLC, it is often the method of choice for samples with many components.

Although separations in HPLC with more than 100,000 plates have been reported, most separations are run with columns that produce 25,000 plates in reverse-phase liquid chromatography (RPLC) mode and less than 10,000 in ion exchange. With CEC, it is possible, with a similar level of difficulty, to achieve about 500,000 plates (RPLC). This improves the chromatographic resolution (R) by

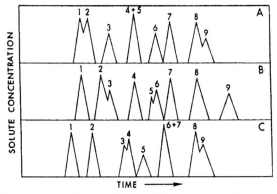

Figure 6. The number of peaks that can fit into a chromatogram is a function of the width of the peaks. In trace B, all the peaks would be separated if the bands were very narrow. This would be possible with very efficient columns. In traces A and C, co-elution is seen for peaks 4+5 and 6+7, respectively. It is possible that much more efficient columns would eliminate the co-elution for these pairs also. (From Ref. [5], with permission.).

$(500,000/25,000)^{0.5}$, or 4.5 times. Thus, a fused peak with an R of 0.3 will appear as two peaks with an R of 1.3, which is almost baseline resolution. In gas chromatography (GC), similar improvements in efficiency of capillary columns compared with packed columns have been a driving force leading to today's substantial preference for capillary columns.

From a practical viewpoint, HPLC instrumentation is very highly developed and optimized. Automation is generally very advanced. Also, the stationary phase in HPLC can be loaded with more sample, avoiding the frustrating overloading effects of CE. CE usually presents trouble when the maximum divided by the minimum sample concentration is larger than 1000. With HPLC, and CEC, the ratio can be larger than 10^5. This means that for HPCE, detection of trace components in a complex mixture often requires two or more runs, which reduces precision and takes time, thereby reducing productivity.

HPCE has the advantage that electricity powers the separation, making it more energy efficient than most HPLC systems. However, CE generally requires charged analytes.

From this introduction, it should be evident that CEC offers the best of both HPLC and HPCE. Since the mobile phase is pumped electroosmotically with CEC, longer separation paths can be used. This increases the number of theoretical plates (efficiency) and, hence, the peak capacity (Table 1).

Table 1. Column Efficiency and Peak Capacity in HPLC and CEC

Particle size (μm)	Length (cm)		Plates per column		Peak capacity	
	HPLC	CEC	HPLC	CEC	HPLC	CEC
5	50	50	55,000	115,000	58	83
3	25	50	45,000	170,000	53	103
1.5	10	50	30,000	250,000	43	125

Adapted from Ref. [3].

The pressure limits of commercial instruments limit the number of plates available in HPLC. In CEC, the flow is generated electroosmotically along the column. Thus, the restriction on length and particle diameter is much less limited. Longer capillaries are possible with CEC, which produces high efficiency (N) and peak capacity (S) compared with HPLC. The peak capacity in Table 1 is calculated from $S = 0.25N^{0.54}$ [6].

If high peak capacity is not required, shorter columns can be used. Assuming that the separation is still adequate, shorter columns reduce run time proportionally to the reduction in column length. To be sure, there are other subtle differences that can be crucial in certain situations, but in general, CEC enjoys the best features of CE and HPLC.

1.2.1 Effect of Pumping System

At first glance, the most obvious difference between HPCE and HPLC is the pumping system. HPLC pumps are usually limited to 6000 psi (40 MPa). This limits the design of columns and ultimately separations. With pumped flow, the flow velocity profile is parabolic. The effect is more problematic with open tubes than with packed beds. A parabolic flow velocity profile reduces the efficiency of the column, since the velocity in the center of any channel can be twice as fast as along the wall. In contrast, the flow velocity profile for CEC is a square wave (Figure 7). Thus, radial mixing or diffusion is eliminated, which doubles the column efficiency (as seen in Table 1 for the 5-μm particles). In addition,

Figure 7. Comparison of flow velocity profiles in open and packed capillaries. (A) molecules diffuse more slowly, which preserves band sharpness. Note the square profile, which has little mixing. (B) Flow velocity profile in an open capillary with pressurized flow. If compounds diffuse radially, as shown by the cross-hatching, more mixing can occur than with A. (C) Flow velocity profile in a packed bed with electroosmotic flow. The square profile maintains the sample in tight bands. Packing imperfections (voids or open spaces) in the column bed do not cause changes in the flow profile. This makes the columns very efficient. (D) Flow velocity profile in a packed bed with pressure-driven flow. The flow is the fastest in regions where the density of particles is low, such as along the wall or in fissures.

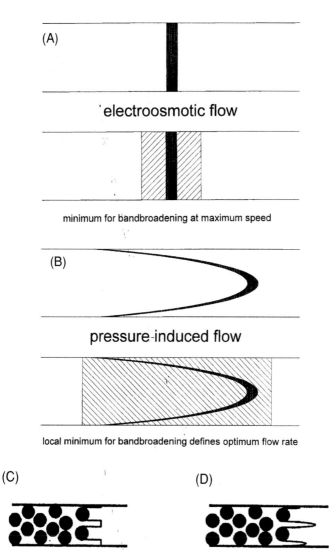

(A)

'electroosmotic flow

minimum for bandbroadening at maximum speed

(B)

pressure-induced flow

local minimum for bandbroadening defines optimum flow rate

(C)   (D)

This facilitates mixing by diffusion in the radial direction. The profile arising from a combination of pressure and electroosmotic flow in a packed bed is the weighted average of the two flows represented by C and D. When the flow arising from pressure drive is low, the band-broadening effect is low. When it is high, it dominates the flow profile. In all cases, the combination of pressure and electroosmotic pumping decreases column efficiency compared with electroosmotic pumping alone. (A and B from Ref. [7], with permission; C and D from Ref. [8], with permission.)

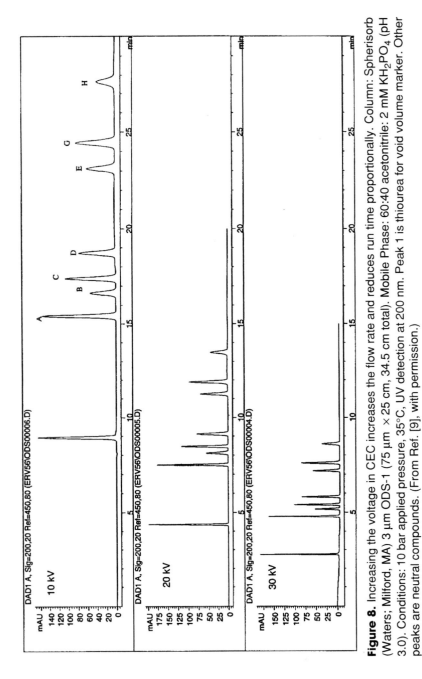

Figure 8. Increasing the voltage in CEC increases the flow rate and reduces run time proportionally. Column: Spherisorb (Waters; Milford, MA) 3 μm ODS-1 (75 μm × 25 cm, 34.5 cm total). Mobile Phase: 60:40 acetonitrile: 2 mM KH_2PO_4 (pH 3.0). Conditions: 10 bar applied pressure, 35°C, UV detection at 200 nm. Peak 1 is thiourea for void volume marker. Other peaks are neutral compounds. (From Ref. [9], with permission.)

electrically driven columns can be much longer than HPLC columns, increasing the resolving power of the technique. The overall effect is that CEC offers a 3- to 10-fold improvement over HPLC. CEC offers improved efficiency over CE, plus the ability to handle samples with a wider range in concentration.

1.2.2 Effect of Voltage

Unlike HPLC, in which the solvent is pumped under pressure, applied voltage is the driving force for CEC. Applied voltage is usually greater than 10,000 V. The effect of voltage on separation speed is illustrated in Figure 8. For a test mix, the retention time of the last peak in the chromatogram at 10,000 V is about 28 minutes [9]. Doubling the voltage cuts the time in half. Tripling the voltage reduces it by 2/3.

What is the effect of voltage on the other parameters that are usually flow rate dependent in HPLC? As shown in Table 2, the number of plates is very high, with optimal efficiency at about 20,000 V. This improves the value of h and R_S but has almost no effect on k'.

1.2.3 Effect of Ionic Strength

Mobile phases in CEC conduct electricity. But what is the effect of the concentration of charges (ionic strength) in the mobile phase? Again, for the sample in Figure 8, increasing the buffer concentration from 2

Table 2. Effect of Voltage on the Chromatographic Parameters for the Last Peak in Figure 8

Parameter	10,000 V	20,000 V	30,000 V
Number of theoretical plates (N)	43,164	58,840	54,560
Reduced plate height (h)	1.93	1.42	1.53
Capacity factor (k')	2.1	2.1	2.2
Resolution (R_s)	6.57	7.73	7.32
u_{ep} (mm/sec)	0.5	1	1.5

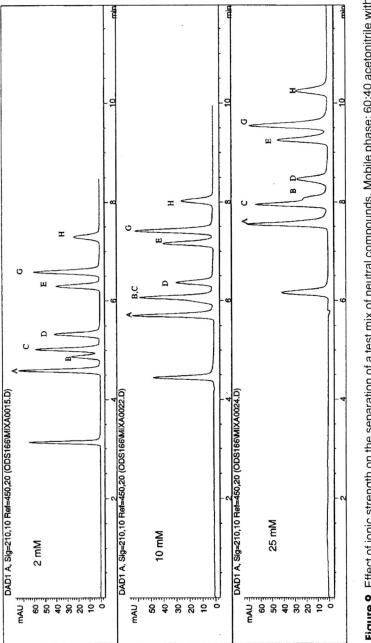

Figure 9. Effect of ionic strength on the separation of a test mix of neutral compounds. Mobile phase: 60:40 acetonitrile with (A) 2 mM, (B) 10 mM, and (C) 25 mM KH$_2$PO$_4$, pH 3.0. Conditions: 30 kV, 10 bar applied pressure; other conditions as in Figure 8. (From Ref. [9], with permission.)

mM to 25 mM increases the elution time for peak H from 7.5 minutes to 10.5 minutes, or about 50%, as shown in Figure 9. Notice that peak B moves much more relative to the rest and eventually reverses elution order with peak C. In general, as the buffer concentration (C) increases, the μ_{eo} decreases according to Equation 3.

$$\mu_{eo} \propto \frac{1}{C^{0.5}} \qquad \text{(Eq. 3)}$$

For this reason, it is common to use a more dilute buffer in CEC than in HPLC; typically the buffer is in the range of 0.5 to 10 mM. In general, one uses the most dilute buffer that still provides useful results.

Mobile phase ionic strength also affects the double layer that describes the transition from the restricted mobility and relatively ordered structure of the liquid–solid surface to the random structure of the

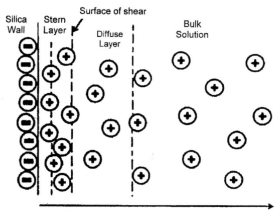

Figure 10. Schematic of the double layer at a charged solid–liquid interface. The Stern layer is nearly immobile, since the opposite charges are strongly interacting. In the shear surface, charges are much more mobile and can move under the influence of an electric field. The orienting effect of the solid decreases across the region from the shear surface to the diffuse layer. In bulk solution, the ions are in random motion. Stronger ionic strength compresses the layers. (From Ref. [2], with permission.)

solution, as shown in Figure 10. Starting with the surface, the negative charges are fixed. To preserve charge neutrality, there is a corresponding layer of cations. These are tightly held in the Stern layer. Even with an applied voltage, they move very slowly. Just above them is another layer of cations that move with the electric field. The boundary is a shear zone in Figure 10. The thickness of the double layer decreases with increasing ionic strength. At 1 mM ionic strength, the double layer is about 8 nm. Knox estimates that particles of a stationary phase should be at least 40 times the diameter of the double-layer thickness—300 nm in this example. Experimental studies show that the optimum is about 1 μm or 1000 nm, which is good verification of a theory with several crude approximations.

1.2.4 Effect of Temperature

The effect of temperature is one of the main differences between HPLC and HPCE. Over a range of about 10 to 20°C, temperature has a very small effect in HPLC. In contrast, most of the irreproducibility of early HPCE separations was quickly traced back to uncontrolled temperature variations causing corresponding changes in the viscosity and electrophoretic mobility. Modern HPCE instruments control the temperature of the capillary to at least ± 0.1°C. For the compounds in Figure 11, using CEC, the effect of temperature on elution time is about 1.3% per °C. Thus, in order for temperature to be an insignificant (<5%) factor in reproducibility measurements for CEC, the temperature should be controlled to ± 0.3°C or better.

Since the elution time is a function of temperature, temperature should be carefully controlled. However, since the plots of retention vs. temperature are usually parallel, there is little advantage of one particular temperature. In general, the temperature is controlled at about 5°C above ambient. Higher temperatures may cause problems due to bubble formation in the buffer.

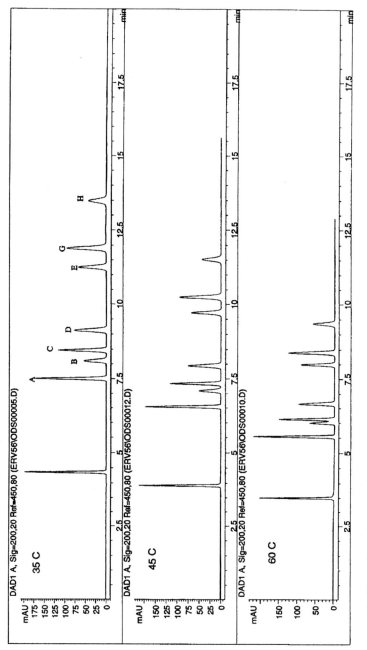

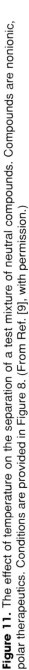

Figure 11. The effect of temperature on the separation of a test mixture of neutral compounds. Compounds are nonionic, polar therapeutics. Conditions are provided in Figure 8. (From Ref. [9], with permission.)

1.2.5   Effect of Mobile Phase Composition

As in HPLC, the mobile phase composition in CEC has a very strong effect on the partition of the sample between the stationary and mobile phases. For RPLC, where the stationary phase is nonpolar, increasing the organic content decreases the retention as shown in Figure 12. As in RPLC, increasing the percentage of organic modifier in CEC mobile phases decreases elution time and resolution. The effect may be reversed at very high levels of organic modifier [10]. Also, at high acetonitrile content, 30 minutes or longer may be required for the current to stabilize at a constant voltage. For 1% reproducibility in retention time, the mobile phase composition should be controlled to about ±0.2%. This is achievable with volumetric glassware. A detailed and specific buffer preparation protocol should be written and followed.

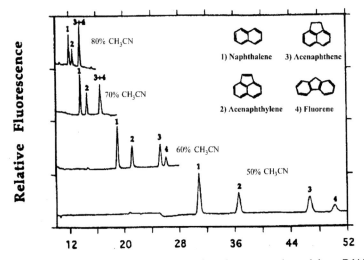

Figure 12. Electrochromatograms showing the separation of four PAHs under different isocratic mobile phase compositions. The column dimensions were 75-μm i.d. × 33-cm packed length. The applied voltage was 15 kV. Injection was performed electrokinetically at 5 kV for 5 sec. The varying concentrations of acetonitrile were in a 4-mM sodium tetraborate mobile phase. (From Ref. [11], with permission.)

1.3 Reproducibility

Scientists have studied extensively the influence of experimental factors for both HPLC and HPCE. This has made possible reproducibility of ± 0.1% to 0.5% in retention time and peak area for both techniques. Achieving this same level of reproducibility is the goal of developers of CEC. With control of the above parameters, it is now possible to

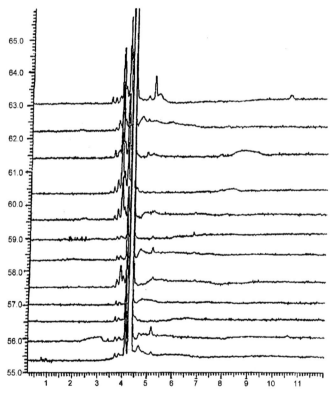

Figure 13. Separation of samples from 12 lots of a drug substance with CEC. The overlay shows that peaks for the major component overlap very well and that the minor peaks are clearly recognizable in each run. The method is used to confirm the identity of the active drug. This was part of a successful development of a validation method for a drug candidate. (From Ref. [12], with permission.)

Table 3. Reproducibility Measures for Chiral Separations

	Run to run (% RSD)	Day to day (% RSD)	Capillary to capillary (% RSD)
μ_{eo}	0.51	1.32	0.68
k'	3.46	3.19	2.24
α	0.53	1.19	0.76

achieve reproducibility, as measured by percent relative standard deviation (% RSD), of better than ± 1%. This is illustrated in Figure 13, which shows the CEC analysis of 12 lots of an active drug substance. Note that the retention times of the major peaks are nearly the same.

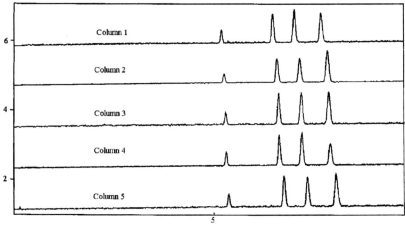

Retention Time (min)

Figure 14. Reproducibility of capillary columns in CEC. Five C_{18} columns: EP-50-30-3-C18 (electrokinetically packed, 50-μm i.d., 30-cm packed length, 3 μm C18). Mobile phase: 80% CH_3CN/20% 4 mM sodium tetraborate. Voltage: 25 kV. Injection: 5 kV/2 sec. Detection: fluorescence, λ_{ex}: 254 nm; λ_{em}: 370 nm. Sample: (1) impurity, (2) fluorene, (3) phenanthrene, and (4) anthracene. (Used with permission of Unimicro Technologies.)

    CEC can provide useful reproducibility, as shown in Table 3 for
chiral separations [13]. The values are about two to five times worse
than obtained using the best of HPLC methods, and about twice as
large as HPCE, but are still quite usable. The electropherograms in
Figure 14 demonstrate that elution time can be controlled very well.
With time and experience, it is probable that the reproducibility will
improve for CEC.

    One significant difference between HPLC and CEC is the stability
of the mobile phase. During CEC elution, hydrogen ions are generated
at the anode, and hydroxide ions are produced at the cathode, as shown
in Figure 15. With time and current, these ions upset the pH of the
buffers, which in turn affects the elution behavior. This explains the
sinusoidal variation in retention times shown in Figure 16, as the buffer
was changed every 10 (Plot A) or 5 runs (Plot B).

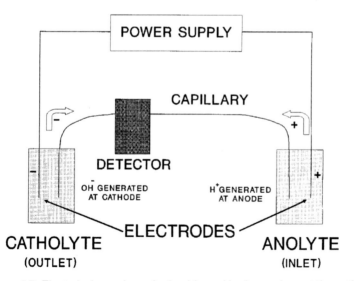

Figure 15. Electrolysis produces hydroxide and hydrogen ions at the cathode and anode, respectively, during CEC runs. These ions affect the pH of
the run buffer. To avoid this problem, the run buffer needs to be changed
periodically. (From Ref. [14], with permission.)

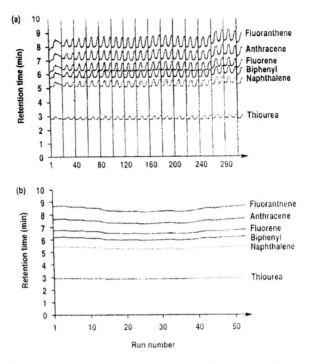

Figure 16. Electrolysis changes the pH of the buffers, as shown in Figure 15. The effect can be minimized by frequently changing the buffer in the reservoirs. Changing every five runs appears to be adequate, at least in this experiment. This is a parameter that needs to be investigated during any method validation study. (From Ref. [3], with permission.)

1.4 Selection of Operating Mode

At the time of this writing, over 90% of the applications of CEC have used reversed-phase columns.

Gradient elution is the one feature of HPLC that is difficult to execute in current HPCE instrumentation. A group at Stanford University (Palo Alto, CA) has developed an elegant design for gradient elution using voltage control (Figures 17 and 18) [15]. The system uses high voltage to produce electroosmotic flow from two reservoirs containing different solvents. In this way it is analogous to a dual pump

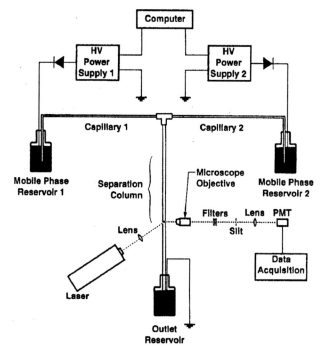

Figure 17. Schematic diagram of a voltage-controlled gradient elution apparatus for CEC. The flow from mobile phase reservoirs 1 and 2 is controlled by the high-voltage power supplies. Since voltage is proportional to the applied voltage, the computer can calculate the fraction of the voltage that each power supply must provide for a specific solvent composition, in the same manner as gradient elution in HPLC with two pumps. (From Ref. [15], with permission.)

gradient HPLC system. The solvent flow from each reservoir is controlled as a function of time, thus varying the overall composition.

1.5 Pressure-Assisted CEC

It may be possible in CEC to increase the flow velocity in the capillary by supplementing electroosmotic flow with pressure pumping. This is called pressure-assisted CEC and is an attractive hybrid for very complex samples.

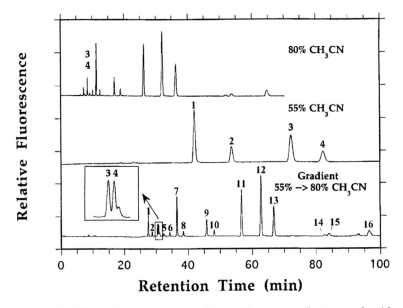

Figure 18. Electrochromatograms of isocratic and gradient runs for 16 polynuclear aromatics of EPA method 610. The applied voltage for isocratic elution was 20 kV. At 55% acetonitrile, only the first four peaks are eluted in less than 90 minutes. At 80% acetonitrile, all compounds are eluted, but the resolution of the early eluters is inadequate. Gradient elution provides good resolution of all peaks shown in the bottom trace. Compounds: (1) naphthalene, (2) acenaphthylene, (3) acenaphthalene, (4) fluorine, (5) phenathrene, (6) anthracene, (7) fluoroanthrene, (8) pyrene, 9) benz[a]anthracene, (10) chrysene, (11) benzo[b]fluoroanthrene, (12) benzo[k]fluoroanthrene, (13) benzo[a]pyrene, (14) dibenz[a,h]anthracene, (15) benzo[ghi]perylene, (16) indeno-[1,2,3-cd]pyrene. (Used with permission of Unimicro Technologies, Inc.)

Shortly after CEC started to attract attention in about 1994, several people, such as Ernst Bayer of Tubingen, Germany, recognized that by adding an HPLC gradient pumping system on the inlet side of the capillary, it was possible to add gradient elution capability to CEC (Figure 19) [16]. Simply adding the pressure drive from the pump enabled the flow rate to be increased beyond that which could be obtained from electroosmotic flow. Electroosmotic flow reaches a limit

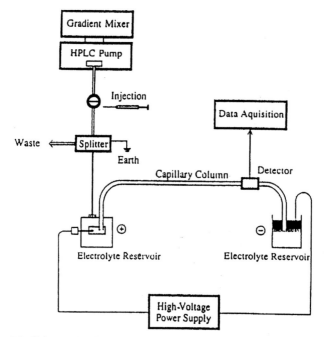

Figure 19. Schematic diagram of electrochromatography system with gradient elution. A conventional HPLC pump feeds the reservoir at the head of the capillary column. If the splitter is pressurized, the pressure from the pump augments the electroosmotic flow. If there is no pressure, then electroosmosis is responsible for flow. (From Ref. [16], with permission.)

because of the breakdown of the insulation of the silica and the limitation of power supplies (usually 30 kV to occasionally 90 kV). However, the flow provided by HPLC pumps is significant only with short capillaries.

Workers at Hewlett-Packard in Waldbronn, Germany [17], approached the subject from the viewpoint of micro (i.e., capillary) HPLC. They achieved a separation with a capillary HPLC system with assistance of 25 kV (Figure 20). It probably makes little difference whether this was pressure assisted CEC (PEC) or electrically assisted capillary HPLC (E-assisted µHPLC). The authors prefer the former,

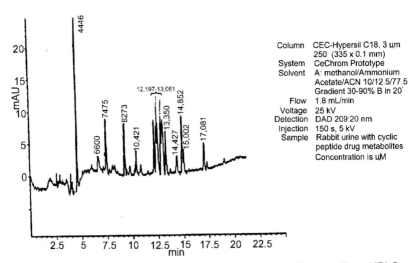

Figure 20. Chromatogram of separation achieved with a capillary HPLC system with assistance of 25 kV. (From Ref. [17], with permission.)

since it operates over a range where pressure assisted flow would make a marginal contribution.

The flow profile offered by the pump is a combination of the square wave of electroosmotic flow and the parabolic flow velocity profile of HPLC shown in Figure 7. This profile sacrifices some of the efficiency inherent in electric pumping for the added speed possible with the combined mechanism. In practice, the combined pumping mode improves performance over HPLC, but the performance is less than with pure CEC.

1.6 Advantages of CEC

Compared with other methods, CEC has several advantages other than the improved efficiency and resolution discussed above.

Perhaps the most important is the mechanical simplicity of the systems. As pointed out by Zare and colleagues at Stanford University, there is no need for mechanical moving parts in a CEC system. With HPLC, pumps and, to a lesser extent, injection valves require maintenance, especially the moving parts, such as piston and injection rotor seals. With CEC, one needs only a power supply. Groups at Oak Ridge [15] and Sandia National Laboratories [18] have described simple fluidic systems that can be etched onto a glass microscope slide. Flow in the eluent and sample flow paths are metered with high voltage, which controls the electroosmotic flow (Figure 21). The design eliminates the need for high-pressure pumps and connections, yet it provides very reproducible separations, even with gradient elution.

Coupling to an MS is a difficult task for HPCE. Low liquid flow and the formation of bubbles are the two most frequent problems. Simply maintaining the proper ground for the MS interface is also a concern. With CEC, Ramsey and Ramsey of the Oak Ridge National Laboratory have used a voltage/flow splitter to feed the majority of the flow into the MS while maintaining a sufficient pressure to prevent bubble formation [19].

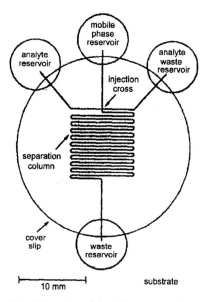

Figure 21. Schematic of the microchip with a serpentine column geometry. (From Ref. [18], with permission.)

CEC uses much less mobile phase than standard column HPLC. A typical HPLC setup running for 8 hours a day uses 100 liters in a year. If it runs full time and during all shifts, the total is about 500 liters. At $20 per liter cost of purchasing and disposal, full-time operation would amount to $10,000 per year per instrument. In contrast, a CEC can run for a year on 200 mL. This may be an important consideration in developing countries and environmentally sensitive locations.

The advantages and disadvantages of CEC compared with HPLC and HPCE are summarized in Table 4.

Table 4. Advantages and Disadvantages of CEC Compared with HPLC and HPCE

Feature	CEC/PEC	HPLC	HPCE
Available theoretical plates	100,000–700,000	<50,000	~200,000
Peak capacity	100+	~50	~100
Mechanical complexity	Simple for CEC; complex for PEC	Complex but well engineered	Less complex than HPLC
Compatibility with mass spectrometers	Better than HPCE	Good	Limited
Mass detection limit	Slightly better than HPLC	Good	Often inadequate
Solvent consumption	<1/100 of HPLC	100 L/year	Similar to CEC
Status of theory	Undeveloped	Mature and useful	Adequate for most purposes
pH operating range	3–11	2–9	2–11
Frits	An experimantal problem	Not a problem	Not a problem
Purchase price ($US)	$30,000 to $50,000	$20,000 to $50,000	$40,000
Cost/run (excludes labor)	$0.02	$2.00	$0.04

1.7 Limitations of CEC

With all the promise of a new technique, there are some limitations with CEC. Some will certainly be overcome, and others may be real constraints.

1.7.1 Operation at Low pH

Many compounds, including peptides and some drugs, are separated by HPLC at pH from 1.5 to 3. Indeed, the pH range of 2.5 to 3 is probably the sweet spot for many HPLC assays. However, the silanols that are responsible for the electroosmotic flow have a pKa near 4, and in turn, the silanols are almost totally protonated (uncharged) at pH < 3. As a result, the electroosmotic flow is nearly wiped out in acidic buffers. One way of getting around this problem is to add alkyl sulfonates (sodium dodecyl sulfonate) to the eluent. This coats the RPLC material with a charge carrier that is strongly ionized at pH 1, and thus provides electroosmotic flow at low pH.

1.7.2 Operation at High pH

Silica column packings account for at least two-thirds of the columns used in HPLC. Depending on the surface chemistry, the bonded phase starts to deteriorate as the pH increases above pH 7. Few silica columns can tolerate pH 10 for any time, limiting their utility at high pH. Polymeric packings might offer greater stability, but few have natural charge carriers, which are required for electroosmotic flow. For this reason, the development of column packings specifically engineered to have a wide pH operating range and useful stability will be an active goal of product development.

1.7.3 Frits

As shown in in Table 1, there is a strong advantage to using capillaries packed with small particles. One study shows, but not conclusively, that the optimal particle for CEC has a diameter in the range of 1.5 to 1.0

μm, not smaller. There is a technical problem in retaining these small particles. Usually, the particles are retained with frits. Up to now, frits have been difficult to fabricate and install, especially reproducibly. Results have been operator dependent.

To get around these problems, several leading researchers are developing capillary columns filled with continuous beds covalently attached to the walls. Covalent attachment immobilizes the stationary phase, and consequently no frit is required. These phases will probably be available commercially very soon. If they live up to expectations and do not introduce other serious liabilities, continuous beds may be a key enabling technology for CEC.

1.7.4 Lack of Theory and Models

Other techniques have developed a set of models and associated theories that are useful in predicting behavior. Some are sufficiently verified to qualify as theory. Today, the understanding of CEC is qualitative at best. Yes, increasing the voltage increases the speed, but the details are missing. This lack of understanding will probably change in less than five years, as the fundamental relationships are developed and verified.

Appendix: Symbols and Units Used in CEC

Symbol	Description	Typical Units/Dimension
μ_{eo}	Electroosmotic mobility	$cm^2\, V^{-1}\, s^{-1}$
U_{eo}	Electroosmotic velocity	$cm\, s^{-1}$
ε_0	Permitivity of vacuum	$8.85 \times 10^{-12}\, C^2\, N^{-1}\, \mu^{-2}$
ε_r	Dielectric constant of the mobile phase (water)	80*
ζ	Zeta potential	40 mV*
η	Viscosity of the mobile phase (water)	$0.008904\, g\, cm^{-1}\, s^{-1}$

Symbol	Description	Typical Units/Dimension
E	Electrical field strength	$V\ cm^{-1}$
σ	Charge density at the surface of shear	$C\ cm^{-2}$
δ	Thickness of electrical double layer	cm
R	Gas constant	$8.3124 \times J\ K^{-1}\ mol^{-1}$
T	Temperature	K
c	Electrolyte concentration	$mol\ cm^{-3}$
F	Faraday constant	$9.65 \times 10^{4}\ C\ mol^{-1}$
ω	Structural parameter related to particle porosity	2.9*
λ	Structural parameter related to flow inequalities in the bed	1.5 for HPLC, 0.1 for CEC
γ	Obstruction factor or tortuosity	0.6
D_m	Diffusion coefficient of solute in the mobile phase	$10^{-5}\ cm^{2}\ s^{-1}$*
d_p	Particle diameter	5, 3, and 1.5 µm
κ	Structural packing diameter	1–15
ψ	Ratio of intraparticulate volume to interstitial volume	0.8
ε_i	Intraparticulate porosity	†
ε_e	Interstitital porosity	†
k	Retention factor	
V_e	Interstitial volume	cm^{3}
V_C	Total column volume	cm^{3}
Ω	Shape factor depending on y interparticulate porosity	†
θ	Tortuosity factor	2
r_c	Channel radius (assumed to be one-third of particle radius)	µm
β	Phase ratio	0.1
k_a	Rate constant for stationary-phase interaction	$1.5 \times 10^{4}\ s^{-1}$
ε_T	Total porosity for the column	0.8*

Symbol	Description	Units/Dimension
ρ_{sil}	Density of solid silica	$2.5 \, g \, cm^{-3}$
ρ_{stat}	Stationary-phase density	$1.0 \, g \, cm^{-3}$
l	Stationary-phase loading	15% (w/w)

*Values that may be different for other systems [3].
†Dimensionless value.

References

1. M. Bier. In *An Introduction to Separation Science*, B. L. Karger, L. R. Snyder, C. S. Horvath, Eds. John Wiley & Sons, New York, 1973, p. 503.

2. M. M. Robson, M. G. Cikalo, P. Myers, M. R. Euerby, and Keith D. Bartle. *Journal of Microcolumn Separations, 9* (5), 358 (1997).

3. M. Dittmann, K. Wienand, F. Beck, and G. P. Rozing. *LC-GC, 13* (10), 803 (1995).

4. J. R. Snyder and J. J. Kirkland. *Introduction to Modern Liquid Chromatography.* New York: John Wiley & Sons, 1979, p. 17.

5. B. L. Karger, L. R. Snyder, C. Horvath. *An Introduction to Separation Science.* New York: John Wiley & Sons, 1973, p. 157.

6. S. Fanali, G. Caponecchi, Z. Aturki. *Journal of Microemulsion Separations, 9*, 9 (1997).

7. A. Manz. Paper presented at CEC Symposium, August 11, 1998, San Francisco, CA.

8. L. A. Colón, K. J. Reynolds, R. A. Maldonado, and A. M. Fermier. *Electrophoresis, 18*, 2162 (1997).

9. P. D. Angus and J. F. Stobaugh. *Electrophoresis, 19*, 2073 (1998).

10. S. E. van den Bosch, S. Heemstra, J. C. Kraak, and H. Poppe. *Journal of Chromatography A*, 755, 165 (1996).

11. C. Yan, R. Dadoo, and R. N. Zare. *Analytical Chemistry*, 68, 2730 (1996).

12. C. Horvath, C. Demarest, N. Smith, P. Angus, and J. Stobaugh. Capillary electrochromatography. Paper presented at HPLC 98, May 3, 1998, St. Louis, MO.

13. C. Wolf, P. L. Spence, W. H. Pirkle, E. M. Derrico, D. M. Cavender, and G. Rozing. *Journal of Chromatography A, 782,* 175 (1995).
14. R. Weinberger. *American Laboratory, 30* (4), 32 (1998).
15. C. Yan, R. Dadoo, R. N. Zare, D. J. Rakestraw, and D. S. Anex. *Analytical Chemistry, 58,* 2726 (1996).
16. B. Behnke, E. Grom, and E. Bayer. *Journal of Chromatography A, 680,* 93 (1994).
17. G. P. Rozing and M. Dittmann. Paper presented at HPCE '98, Orlando, FL, January, 1998.
18. S. Jacobsen, R. Hergenröder, L. Koputny, and J. M. Ramsey. *Analytical Chemistry, 66,* 2369 (1994).
19. R. S. Ramsey and J. M. Ramsey. *Analytical Chemistry, 69,* 1174 (1997).

2 Capillary Column Technology

Column design is still the weakest segment of CEC technology. Initially, researchers at the forefront of CEC technology felt that simple HPLC column packings would work well. They reasoned that, after all, capillary electrochromatography was just a different type of pumping system on capillary columns. That dream evaporated in less than a year. Today, it is clear that capillaries packed with HPLC packing are only an interesting starting point. Realists recognize that a long development cycle has only just begun.

The initial euphoria and optimism about CEC can be attributed to some astounding successes with capillaries packed with common HPLC packings. The nature of electroosmotic flow made even poorly packed columns look much better than most investigators had seen with HPLC. They jumped to the podium from all over the world demonstrating very satisfying results, such as those shown by Rozing and Dittmann in Figure 1 [1].

It is interesting that the fastest results were obtained on the older column packings. This success is attributed to the presence of silanols in the older stationary phases developed for HPLC.

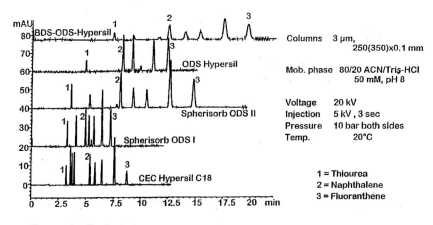

Figure 1. Typical chromatograms of polyaromatic hydrocarbons (PAHs) obtained with capillaries packed with commonly available RPLC particles. The top chromatogram (BDS-ODS-Hypersil) shows a retention time that is longer due to the low number of silanols. ODS Hypersil, in contrast, has more silanols and shorter retention times. Spherisorb ODS II is end capped, which removes silanols. Spherisorb ODS I is not end capped, and hence is much faster. CEC Hypersil C-18 is not end capped, which provides residual silanols and promotes electroosmotic flow. (From Ref. [1], with permission.)

For a while, there was some controversy about the charged groups on the surface of the capillary. However, Horvath has found that the column packing was responsible for instigating most, if not all, of the electroosmotic flow [2].

After the initial success, real-world problems set in. There is a difference between do-it-once-and-publish results and doing it daily with the same results, which revealed recurring problems. It was found that the capillaries broke easily, especially at the detection window, which was free of polyimide coating. Capillary-to-capillary reproducibility was poor by contemporary HPLC standards. Manufacturers found that the yield of marketable capillaries was low. Capillary longevity was too short for useful work. Of course, the capillaries were being evaluated against the performance standards of state-of-the-art HPLC columns, which had evolved over 25 years. Most of the complaints had a familiar ring, since they have been heard before for GC, HPLC, and HPCE. These problems will likely be solved with time.

However, there was something new in the list of complaints: frits! In HPLC columns, sintered metal plugs are cut to form frits fashioned into the end fitting. But in fused silica capillaries, there is no endfitting. Plus, the column packing is often smaller than is practical for HPLC. Frit technology has not yet transferred from HPLC.

Fortunately for CEC, there have been several good chemists working on monolithic supports or continuous beds for HPLC [3–11]. Some of these are actually bonded to the wall, which eliminates the need for any frit. Some supports have interesting chemistry that should allow them to be exploited to improve selectivity. However, they have not yet been developed into products. *Monoliths* seems to be the more popular term now, but *continuous beds* was used by Hjerten in 1989 to describe this technology [29].

2.1 Packed Capillaries

Even if the frit problem could eventually be solved, it quickly became apparent that the optimal column packing for CEC was not the same as for HPLC.

2.1.1   Particle Size

The resistance to Poiseuille flow effectively limits the particle diameter in HPLC to something in the range of 1.5 to 3 microns. Below this size, the columns require pressure that is higher than most HPLC pumps can deliver. In contrast, CEC uses electroosmosis to pump the liquid, which allows one to use particles smaller than one micron. This has been investigated by Lüdtke working in the laboratory of Unger of the University of Johannes Gutenberg University (Mainz, Germany). Today, it appears that 1 µm is about the optimum [12]. However, this may be an experimental artifact, since the CEC system being evaluated was not optimized for measuring smaller peak volumes expected by still smaller particles. There is reason to believe that there is a minimum particle size useful for CEC since the thickness of the double layer should be much smaller than the particle diameter. It is expected that if the double layer thickness is similar to the particle diameter, overlap will occur, which would cause a large decrease in electroosmotic flow. However, this has yet to be confirmed experimentally.

The difficulty in finding the optimal particle size for CEC is related to the additive nature of the peak-broadening process of chromatography. With the current lack of an extensive body of literature on CEC, it is difficult to separate the peak broadening arising from the instrument (called extra column effects) from the broadening of the column itself.

Even so, it is possible to make a capillary with particles as small as 1 µm. These packings provide a reduced plate height of 1 µm, or a million plates per meter! But how long can the capillary be? The practical limit today is about 20 to 25 cm. However, some have used longer capillaries, which provide proportionately greater plate counts. These provide spectacular separations until one looks at the time. Still, these capillaries provide separations that simply are not practical by HPLC.

2.1.2 Pore Structure

Pore diameter appears to be important, as shown in Figure 2. The optimum for several small molecules is about 100 to 300 Å. Yan, Demarest, and others have investigated nonporous particles for CEC (Unimicro Technologies, Pleasanton, CA). These packings show very high efficiency, attributable primarily to the small particle size. It is probable that packed capillaries with the highest number of plates will be constructed from nonporous particles with bonded ionic groups. These capillaries appear to be even more efficient than the monoliths described in Section 2.5.

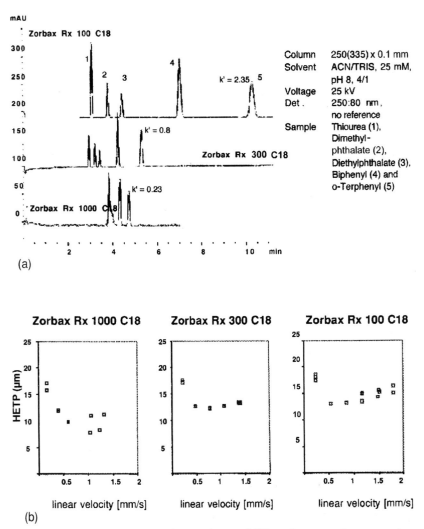

Figure 2. (a) Comparison of pore size on CEC performance in reversed-phase CEC. Large pore materials appear to lack the capacity to provide good separation. With 100- or 300-nm pores, the capacity and kinetics seem to be favorable. (b) HETP vs. linear velocity for RP phases with different pore size. (From Ref. [13], with permission.)

2.2 Surface Chemistry on Particles

2.2.1 RPLC

As in HPLC, RPLC packings are the most common for CEC. To be effective in providing electroosmotic flow, these packings need to have charge-carrying groups. If silanols are used as charge carriers, the weak acid strength decreases the electroosmotic flow as the pH drops below about 4.5. Inability to operate at pH 3 is a serious impediment to converting many methods from RPLC to CEC.

To get around the problem of transferring methods run at low pH, some have adsorbed or bonded ionic groups such as alkyl sulfonates to the silica, as illustrated in Figure 3 [14]. El Rassi (Oklahoma State University, Stillwater, OK) devised a three-layer approach for open tubular and packed capillaries. The first layer was made by attaching γ-glycidoxypropyltrimethoxysilane on the surface of beads. Next, alkene sulfonates were grafted, followed by dimethyloctadecylchlorosilane, to produce octadecylsulfonated silica (ODSS). The resulting stationary phases show 20% to 40% higher electroosmotic velocity than the simple ODS packing [14].

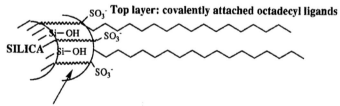

Figure 3. Layered structure for ODSS stationary phase for CEC. The silica is encapsulated in a glycidyl layer to which sulfonic acid groups are attached to provide charge carriers. The outer surface is octadecyl chains to provide reversed-phase retention as in HPLC. (From Ref. [14], with permission.)

2.2.2 Ion Exchange

Given the success of RP-CEC, the extension of CEC to other modes that have been developed using HPLC is obvious. But some of the results, for example for cation exchange, have been quite unexpected. Smith (Imperial College, London, UK) found that some injections produced exceptionally narrow peaks [15]. The efficiency was over 50 million plates per meter, or about 1000 times higher than observed in the best HPLC ion exchange columns. However, run-to-run reproducibility was poor, even with the same sample and mobile phases. The mechanism producing the spectacular results is not understood. For this reason, ion exchange CEC is the subject of intense research, but not much is being done in applied laboratories.

2.2.3 Other Modes

As of this writing, there are scattered preliminary reports of using CEC in modes analogous to GPC, HIC, and normal-phase HPLC. These results have been presented in lectures, but the results are preliminary and beyond the scope of this book.

2.3 Techniques for Packing Capillaries

2.3.1 Capillary Geometry

The design of CEC instrumentation remains far from optimized, particularly for injection and detection. Capillaries for CEC often have constraints placed on them by the design of the instruments that make capillary fabrication more difficult than for HPLC. Most examples in the literature have dead spaces or transfer tubes. Optical detectors require fixed positions for the flow cell, which is not near the end of the capillary. It is hoped that improvements in commercially designed instruments will improve the practicality of the instrumentation.

2.3.2 Capillary Length and Cutting

The capillary length is important in achieving the separation: too short, and the resolution is inadequate; too long, and the run time is too long, which lowers productivity. One of the most interesting aspects of the monolithic capillaries, described in Section 2.5, is the ability to make very long capillaries and then cut off just the length that one requires for a separation. This is illustrated for a series of test compounds in Figure 4. Increasing the active length of the capillary increases both resolution and run time, yet the overall pattern remains the same.

Cutting capillaries to length is no problem if the right technique is used. A small fused silica cutter from Agilent Technologies (Palo Alto, CA), is perhaps the most convenient. With a little practice, a silica cutter produces a quick and square cut. The ends of any cut should be routinely examined under a microscope (at least 60×) to confirm the geometry of the cut. Optical fiber cutters are also useful and more reliable but are significantly more expensive. Failure to have a square end will cause problems in peak shape and irreproducibility in electrokinetic injection.

Length:

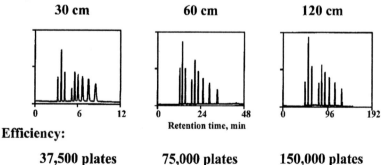

| 30 cm | 60 cm | 120 cm |

Efficiency:

| 37,500 plates | 75,000 plates | 150,000 plates |

Figure 4. Doubling of column length doubles run time and capillary efficiency. The capillaries are filled with a porous polymer that is polymerized in place. One can cut off the length that is required for any separation. (From Ref. [16], with permission.)

2.3.3 Frit Formation

Packed capillaries rely on frits to retain the stationary phase. Frits must be small and not add to the band broadening from the column. Furthermore, they may need to be formed in the interior of the capillary, and in the presence of the stationary phase, as illustrated in Figure 5.

There are several protocols for forming frits. One uses a paste of silica and sodium silicate, which is heated to 400ºC or above. (The specific temperature is determined empirically.) This is particularly good for frits formed at the end of the capillary. Another uses a heated wire to thermally fuse the column packing. With a heated wire, frits can be formed any place along the capillary, which is a definite advantage over the silicate paste protocol.

2.3.4 Window Formation

Optical detectors, such as UV, diode array, and fluorescence, perform better if the polyimide coating on the surface of the capillary is removed. The coating is designed to protect the silica from stress fracturing caused by adsorption of traces of water. Without the coating, the fused silica is very fragile and can break easily, sometimes apparently spontaneously.

Some recommend removing the coating mechanically with a knife. The success of this approach is operator dependent. Others burn a hole with electrically heated coils, as in Figure 5, or a flame from a microtorch. The best results appear to be from fixing the capillary in a narrow groove in flat glass or stainless steel. A small drop of concentrated sulfuric acid is placed on the section that needs to be clear. After the acid is washed away, one has a fragile but clear window. Recently, UV transparent coatings have been introduced to replace conventional polyimide (Supelco Corp., Bellefonte, PA). This coating may eliminate the need for this bit of art.

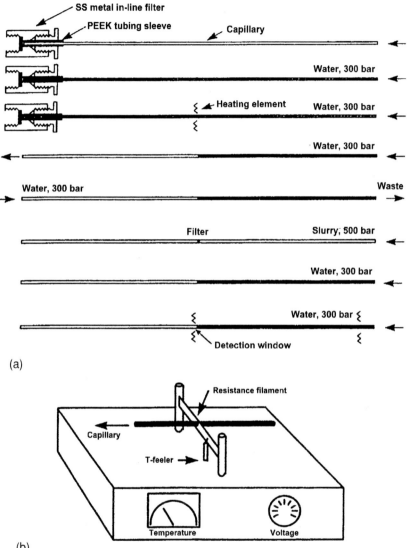

Figure 5. The steps involved in producing packed capillaries for UV or fluorescence detection. For practical reasons, the detection window and the retaining frit are fabricated away from the ends of the tube. Heating to specific temperature forms the window and the frit. The power supply is shown at the bottom. (From Ref. [17], with permission.)

2.4 Packing Protocols

Packing the capillaries is time consuming and costly. As shown in Figure 5, packing the capillary involves several steps: packing, frit formation, and unpacking, which must be performed in the proper sequence [18]. Starting from the top, the capillary is washed, then filled with packing such as bare silica for formation of the frit (SS metal in-line filter) with the heating element. The excess column packing is washed out from either end, leaving the filter or frit in the interior. Next, the capillary is packed with a RPLC packing. The detection window is burned off in the last step. The temperature for formation of the frit and window is measured and controlled with the power supply (T-feeler).

The exact protocol for packing the slurry varies greatly. Bartle (University of Leeds, Leeds, UK) and others have investigated this [18]. Bartle compared supercritical fluid with conventional slurry packing for six capillaries packed with four methods, as shown in Table 1 and Figure 6. The median performance values did not vary much, but the standard deviations did. For slurry packing, the large %RSD leads one to anticipate the need to test each capillary and to anticipate a poor yield of useful columns. A wide range of efficiencies can lead to problems, since a separation developed on an unusually efficient column may be unsatisfactory when run on a capillary with only average efficiency.

One of the more interesting techniques has been developed in the laboratory of Colón at the State University of New York at Buffalo [19]. Colón's lab constructed a centrifuge with the slurry reservoir in the center. The capillary rotates around the reservoir, much as the spokes of a bicycle wheel. This packs the capillaries very well. A tractor tire casing was used as the scatter shield. An electrochromatogram is shown in Figure 7.

2.4.1 Electropacked Capillaries

Taking advantage of the ability of the particles to move in an electric field, Unimicro Technologies (Pleasanton, CA) packs the capillaries

Table 1 Comparison of Packing Methods

	Migration time (min)			Electroosmotic flow (mm s^{-1})			Efficiency (plates m^{-1})		
	Mean	SD	% RSD	Mean	SD	% RSD	Mean	SD	% RSD
CO_2	4.17	0.42	10.15	1.65	0.14	8.46	208,130	28,430	13.66
CO_2 + probe	3.73	0.38	10.32	1.74	0.99	5.03	199,210	28,385	14.30
Slurry	3.99	0.87	21.88	1.61	0.22	14.0	219,860	58,680	26.69
Slurry + probe	3.69	0.20	5.42	1.69	0.12	7.31	213,880	26,987	12.57

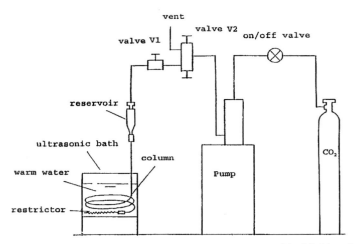

Figure 6. Apparatus for packing capillaries using supercritical fluid carbon dioxide and slurries. The slurry or suspension of column packing is placed in a reservoir. The capillary is immersed in water, with or without ultrasound agitation. The pump displaces the slurry into the capillary in a manner similar to packing HPLC columns. (From Keith Bartle, with permission.)

using electroosmotic flow. This enables longer packing lengths with high uniformity, since one is not limited by the applied pressure. Furthermore, the particles are packed with the same forces that they will experience in use. This may avoid problems with mechanical shock during packing the first part of the column.

2.4.2   Capillary Evaluation

After packing, each capillary should be tested prior to use (Table 1). The testing should include a marker for electroosmotic flow. Thiourea is the most frequently used test probe and should provide a RSD of about 2% from column to column. The packing efficiency can be evaluated by using a retained peak such as biphenyl. The reduced plate height should be less than 2.0 and the asymmetry close to 1.0. One can also use another hydrocarbon and measure the selectivity. The RSD on migration times should be no worse than 4%.

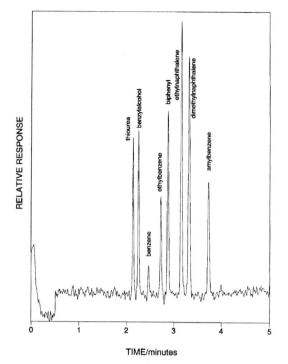

Figure 7. Electrochromatogram for the separation of a test mixture on a capillary column packed by centripetal forces. Column packed at 1500 rpm with 3 μm ODS particles in a slurry solution of methanol. Column was pressurized (5000 psi) for washing with acetone after packing. Separation conditions: fused silica capillary 30 μm i.d. × 27 cm length (20 cm to detector), 2 s injection at 1 kV; mobile phase: acetonitrile/4 mM borate buffer, pH = 9 (80/20). Runs were performed in a commercial CE unit (P/ACE 2200, Beckman) at a temperature of 15°C without pressurization. (From Ref. [20], with permission.)

2.4.3 Capillary Failure

Capillaries will eventually fail and must be replaced. In 1998, some scientists achieved as many as a thousand injections, but a few hundred is far more common. Breakage near the window is the most common failure mode. There is no known recovery from this.

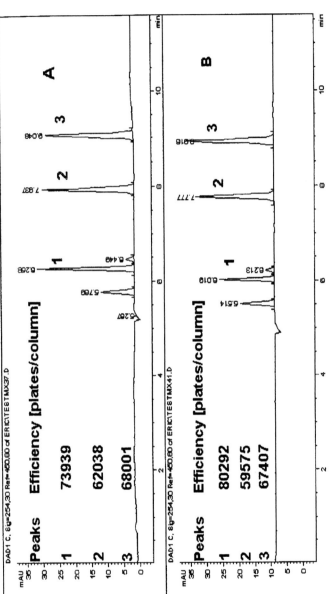

Figure 8. Capillary performance (A) before and (B) after capillary "repair." Performance conditions of capillary before voiding: 25 kV; mobile phase, 75% CH$_3$CN, 25% 50 mM/L Tris buffer (pH adjusted to 7.5 with HCl); 254 nm; 15°C; length, 30 cm. Peaks (in order of elution): thiourea, biphenyl, benzamide, anisole, benzophenone. Capillary length after voiding and subsequent repair: 25 cm. All efficiency figures are rounded to the nearest thousand. Capacity factors are calculated using the relationship K′ = t$_0$/t$_0$ and the solvent front as the t$_0$ marker. (From Ref. [20], with permission.)

All too often, failure results from the settling of the column packing, creating a void, common in HPLC columns. Settling usually occurs in the first few runs after installing a new column. Rather than simply replacing the capillary, one can let the flow go for a few hours to stabilize the settling and then cut off the empty section, form a new frit, and retest the column. If performance is satisfactory, then the column will probably provide good service A successful regeneration of a CEC capillary is illustrated in Figure 8.

Slow contamination of the column packing by materials from the instrument, mobile phase, and sample may result in a slow deterioration of performance in terms of efficiency, symmetry, or resolution. To some extent, deterioration is unavoidable, but if the capillary lifetime is too short, or the expense is high, then it may be worthwhile to systematically investigate the possible source(s) of the problems and develop steps to eliminate them. Although troubleshooting protocols have been worked out for GC, HPLC, and HPCE, there is not yet enough experience in CEC. Until these protocols are established, one should review the causes of column failure in HPLC for clues to the possible cause. Capillary vendors are also a source of advice and help.

2.5 Continuous Beds or Monoliths in CEC

The technology associated with continuous bed, or monolithic, columns and the extensions into highly specific separations continue to be explored.

Compared with packed capillaries, monolithic capillaries are much easier to prepare. For RPLC, Svec (University of California, Berkeley, CA) starts with a mixture of butylmethacrylate (for hydrophobic retention), ethylene dimethacrylate (a cross-linker), and 2-acrylamido-2-methyl-1-propanesulfonic acid (AMPS), which provides a charge carrier for electroosmotic flow [16, 21, 22]. To control the porosity, 1-propanol, water and butanediol are added. The capillary is filled with the polymer cocktail with a simple syringe. The capillary is sealed and placed in a hot bath, which initiates the polymerization

1) FILLING   2) POLYMERIZATION   3) WASHING

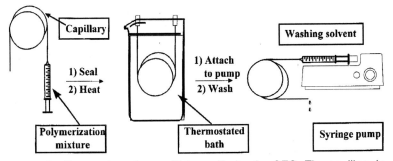

Figure 9. Preparation of monolithic capillaries for CEC. The capillary is filled with the prepolymer cocktail, including porogen (a solvent used to control the porosity of the final polymer) plus monomers, to produce hydrophilic retention, cross-linking, and charge carriers. Polymerization is thermally initiated and controlled. Finally, the porogens are flushed out with a syringe pump and the capillary is ready for conditioning, testing, and use. (From Frantisek Svec, with permission.)

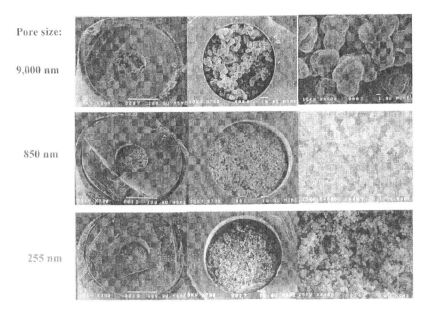

Figure 10. Beds polymerized in capillary tubes. The pores vary depending on the composition of the mixture of monomers and porogens. (From Frantisek Svec, with permission.)

reaction. After a few hours, the capillary is removed and washed with a syringe pump to remove the porogens, as shown in Figure 9.

The morphology of the monolith can be varied over a wide range. Pore size is perhaps the most important parameter. Figure 10 compares the difference in morphology from 9000 nm to 850 nm to 255 nm. The monolith in each case is made up of congealed globular structures. The dimensions are a function of the percentage of solids and the feature size of the aggregate. The chromatograms in Figure 11 show that pores of about 600 nm provide better resolution than the larger pores for the test compounds, which are all small.

Svec, Remcho, and colleagues have explored molecular imprinting to make polymers with chiral selectivity [21–25]. Molecular imprinting involves polymerization of a stationary phase in the presence

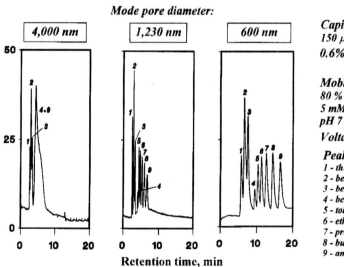

Mode pore diameter:

| 4,000 nm | 1,230 nm | 600 nm |

Capillary:
150 μm i.d. x 30 cm
0.6% AMPS

Mobile phase:
80 % acetonitrile in
5 mM phosphate,
pH 7

Voltage: 25 kV

Peaks:
1 - thiourea
2 - benzyl alcohol
3 - benzaldehyde
4 - benzene
5 - toluene
6 - ethylbenzene
7 - propylbenzene
8 - butylbenzene
9 - amylbenzene

Retention time, min

Figure 11. Chromatograms obtained with columns shown in Figure 10. Chiral separations are by molecular imprinting. (From Frantisek Svec, with permission.)

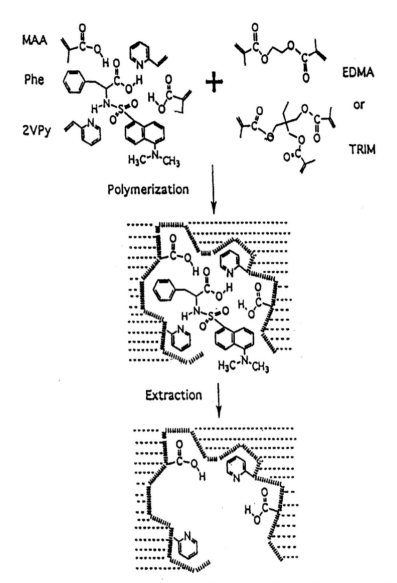

Figure 12. Molecular imprinting for chromatographic stationary phase in CEC. A mixture of analyte, monomers, cross-linkers, and initiators is pumped into the capillary and heated to initiate the polymerization. The analyte is washed out, leaving a cavity that is selective for the analyte. This provides a highly selective stationary phase. (From Ref. [24], with permission.)

of an analyte. This is illustrated in Figure 12. A mixture of monomers (olefins) and cross-linkers (polyolefins) is polymerized in the presence of an analyte such as dansyl phenylalanine [23]. The analyte is washed away, which leaves a cavity that has the functional groups locked in a rigid geometry. Assuming that the monomer and analyte were in a low-energy configuration prior to polymerization, it is likely that they will be in a similar low-energy but locally ordered configuration after polymerization. The polymerization can be thought of as making a casting of an object such as a figurine by forming plaster around the object. Once the plaster sets, the object can be removed and the mold retains its shape selectivity. Only objects with strong shape similarity are able to interact with the surface in the cavity of the mold. This provides lock-and-key selectivity. However, the analytes must not contain functional groups that will participate in or inhibit the polymerization reaction.

Svec's group has developed monolithic columns using a copolymerization approach to provide separation of enantiomers of N-(3,5-dinitrobenzoyl)leucine diallylamide, as shown in Figure 13 [25].

2.6 Open Tubular Capillaries

In GC, most separations today are performed on capillary columns. Ideally, the stationary phase is present in a uniform, thin film on the fire-polished surface of the capillary. In some cases, the quantity of stationary phase available on the wall is not sufficient to yield a useful separation. Rather than revert to a packed column, which has a much higher proportion of stationary phases than a capillary, some chromatographers use open tubular capillaries with active sorbents embedded along the capillary wall.

Pesek at San Jose State University (San Jose, CA) has developed analogous capillaries for CEC [26]. Starting with a thin, 20-μm internal diameter capillary and etching the surface with caustic reagents, he produced a variety of surface structures ranging from ripples to coral-like rods. He then added a bonded stationary phase to produce open tubular capillaries (OTCs), reversed-phase capillaries for CEC. These

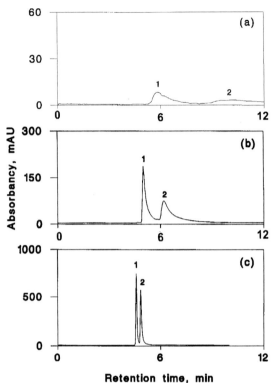

Figure 13. Separation of enantiomers of *N*-(3,5-dinitrobenzoyl)leucine di-allylamide with CEC with a copolymerized capillary stationary phase. The stationary phase was prepared by in situ polymerization of a mixture of alkylacrylates and a chiral monomer in the capillary. Electroosmotic flow is provided by 2-acryl-amido-2-methylpropane-1-sulfonic acid in the polymerization mixture. (From Ref. [25], with permission.)

capillaries have been shown to have smaller plate heights due to the lack of band-broadening effects associated with the existence of packing materials and end-column frits. High concentration sensitivity is another advantage of OTCs, since columns with extremely small internal diameter (25 µm) are used. Also, the small diameters of the OTCs allow for the use of a higher voltage in CEC without significant Joule heating temperature effects.

Pesek has also developed a novel approach for bonding of organic ligands to a silica surface [27]. Conventional bonding of organic ligands to silica has usually involved a siloxane (Si-O-Si) bond. Pesek's approach utilizes a much more pH- and aqueous-stable Si-C bond.

$$\geqq Si - OH + (OEt)_3SiH \longrightarrow \ \geqq Si - O - Si - H + 3 \ EtOH \qquad (Eq. \ 1a)$$

$$\geqq SiO - Si - H + R - CH = CH_2 \longrightarrow \ \geqq Si - O - Si - CH_2 - CH_2 - R$$
$$(Eq. \ 1b)$$

The fused silica inner surface was etched with ammonium hydrogen difluoride to increase (up to 10×) the surface area. The chemistry used to modify the etched capillary was based on a silation–hydrosilation reaction scheme, which leads to a direct silicon–carbon bond on the surface.

In this process (Equations 1a and 1b), the etched surface of the capillary was first reacted with triethoxysilane (TES) to produce a hydride surface. An organic moiety was then attached to the hydride intermediate by passing a solution containing a terminal olefin and a suitable catalyst, such as hexachloroplatinic acid, through the capillary.

A comparison of the separation characteristics for a series of lysozymes on bare, unetched, diol-modified–etched and C18-modified–etched capillaries is shown in Figure 14 [28]. Figure 15 illustrates the surface of a typical etched fused silica capillary [28]. For each column, the elution order was the same, indicating that while solute-bonded phase interactions may have been significant, the differences in electrophoretic mobility were primarily responsible for the separations observed with the angiotensins.

2.7 Conclusions and Prognostications

The preceding discussion presents the highlights and problems of current developments in CEC capillaries. Each approach (packed, monoliths, or open tubular) offers particular advantages along with developmental problems. Monoliths offer the lowest cost and most flexibility, since one can cut them to length. Packed capillaries probably offer the highest efficiency (plates/meter) but are also the most expensive to manufacture. The optimal particle diameter is in the range of 0.5 to 1.0 µm. Open tubular CEC may be a good compromise between the other two designs.

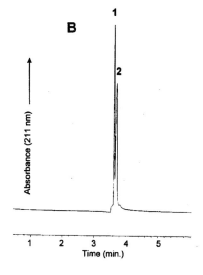

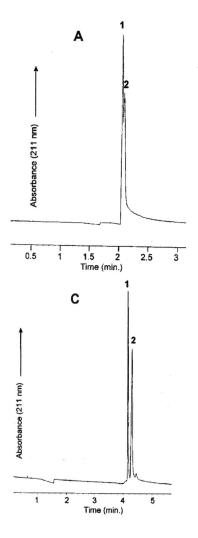

Figure 14. Separation of chicken and turkey lysozymes on 20 µm i.d. (A) bare capillary, (B) unetched C18-modified capillary, and (C) etched C18-modified capillary. Conditions: V = 30 kV, pH = 3.7, detection at 211 nm, injection 2 s at 12.5 cm Hg vacuum (1 mm Hg = 133.332 Pa), length = 50 cm (A and B), 51.5 cm (C), 1 = 25 cm (A and B), 22 cm (C). Solutes: 1 = turkey, 2 = chicken. (From Ref. [28], with permission.)

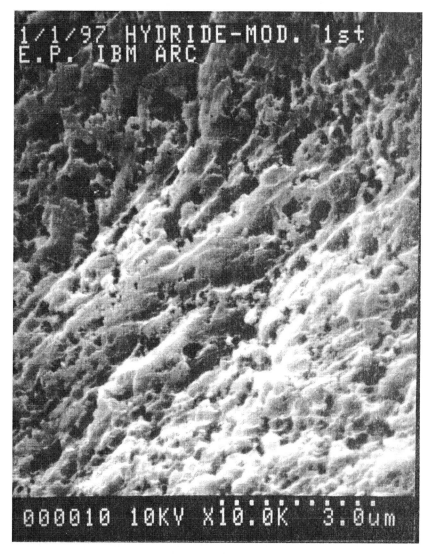

Figure 15. Scanning electron micrograph of 20 μm i.d. etched fused silica capillaries: high-magnification image obtained for capillary etched at 300°C for 2 h followed by 400°C for 2 h. (From Ref. [28], with permission.)

References

1. M. M. Dittmann and G. P. Rozing, *Journal of Chromatography A,* *744*, 63 (1996).
2. G. Choudhary and C. Horvath, *Journal of Chromatography A, 781,* 161 (1997).
3. C. Fujimoto, *Analytical Chemistry, 67,* 2050 (1995).
4. C. Fujimoto, J. Kino, and H. J. Sawada, *Journal of Chromatography A, 716,* 107 (1995).
5. C. Fujimoto, Y. Fujise, and E. Matsuzawa, *Analytical Chemistry, 68,* 2753 (1996).
6. L. Schweitz, L. Andersson, and S. Nilsson, *Analytical Chemistry, 69,* 1179 (1997).
7. A. Palm and M. V. Novotny, *Analytical Chemistry, 69,* 4499 (1997) .
8. S. Hjerten, D. Eaker, K. Elenbring, C. Erickson, K. Kubo, J.-L. Liao, C.-M. Zeng, P.-A. Lindstrom, C. Lindh, A. Palm, T. Srichiayo, L. Valtcheva, and R. Zhang, *Japanese Journal of Electrophoresis, 39,* 105 (1995).
9. J.-L. Liao, N. Chen, C. Erickson, and S. Hjerten, *Analytical Chemistry, 68,* 3468 (1996).
10. C. Erickson, J.-L. Liao, K. Nakazato, and S. Hjerten, *Journal of. Chromatography A, 767,* 33 (1997).
11. E. C. Peters, M. Petro, F. Svec, and J. Frechet, *Analytical Chemistry, 69,* 3646 (1997)
12. S. Lüdtke, T. Adam, and K. K. Unger, Towards the ultimate minimum particle size in electrochromatography, Lecture at HPCE 98, Orlando, FL, 1998.
13. M. M. Dittmann and G. P. Rozing, Capillary Electrochromatography (CEC): From a Theoretical Concept to a Practical Applications. Presented at the 20th International Symposium on Capillary Chromatography, Riva del Garda, Italy, 1998.
14. M. Zhang and Z. El Rassi, *Electrophoresis, 19,* 2068 (1998).
15. N. W. Smith and M. B. Evans, *Chromatographia, 41,* 197 (1995).
16. F. Svec, Monolithic capillaries for electrochromatography, Paper presented at HPLC 98, St. Louis, MO.
17. S. E. van den Bosch, S. Heemstra, J. C. Kraak, and H. Poppe, *Journal of Chromatography A, 755,* 165 (1966).

18. S. Roulin, K. D. Bartle, P. Myers, M. R. Euerby, and C. Johnson, Unpublished results.

19. L. A. Colón, K. J. Reynolds, R. Alicea-Malsonado, and A. M. Femier, *Electrophoresis, 18*, 2162 (1997).

20. R. J. Boughtflower, T. Underwood, and J. Maddin, *Chromatographia, 41*, 7/8, 398 (1995).

21. E. C. Peters, Petro, F. Svec, and J. M. Frechet, *Analytical Chemistry, 70*, 2288 (1998).

22. E. C. Peters, M. Petro, F. Svec, and J. M. Frechet, *Analytical Chemistry, 70*, 2288 (1998).

23. Z. J. Tan and V. T. Remcho, *Electrophoresis, 19*, 2055 (1998).

24. V. Remcho, *Electrophoresis, 19*, 1055 (1998).

25. E. Peters, K. Lewandowskis, M. Petro, F. Svec, and J. M. Fréchet, *Analytical Communications, 35*, 83 (1998).

26. J. J. Pesek, M. T. Matyska, J. E. Sandoval, and W. J. Williamson, *Journal of Liquid Chromatography and Related Technologies, 19*, 2843 (1996).

27. J. J. Pesek, M. T. Matyska, W. J. Williamson, M. Evanchic, V. Hazari, K. Konjuh, S. Takhar, and R. Tranchina, *Journal of Chromatography A, 786*, 219 (1997).

28. J. J. Pesek and Matyska, *Journal of Chromatography A, 736*, 255 (1996).

29. S. Hjerten, et al., *Journal of Chromatography, 473*, 273 (1989).

3 The Mobile Phase

Compared with the mobile phase in HPLC, the mobile phase in CEC has received much less attention. Current practice is for similar mobile phases to be used for both techniques, but it is probable that the CEC mobile phase can be improved upon. The mobile phase plays a dual role in CEC. It is the eluent, and in combination with the stationary phase, it is also the charge carrier producing electroosmotic flow. Thus, parameters such as ionic strength, conductivity, Joule heating, and double-layer thickness, and viscosityall of which are functions of composition and temperature of the mobile phaseare much more important in CEC than in HPLC.

Achieving reproducible separations in CEC requires the same careful techniques found in HPLC and HPCE. Mobile phases need to be prepared the same way every time. Too often, the column is blamed for irreproducibility, when sloppy preparation of the mobile phase is the primary cause. For example, overshooting on the adjustment of pH and then back-titrating can increase the ionic strength, which will change the electroosmotic flow, and hence the retention time.

The following observations designed to improve interlaboratory transfer of HPCE methods may be applied to CEC, where the mobile phase serves much the same purpose.

> The composition of a solution must be given in such a way that the solution can be prepared without any doubts. It is preferable to indicate the composition of a solution by giving the molarities (or mass) of all components present. To indicate that a solution was titrated to a certain pH is not sufficient because this procedure may include other effects, e.g., dissolved CO_2 from air. In that case it is advisable to give the experimentally measured pH of the prepared solution. The preparation of all solutions should be specified, e.g., not only the running buffer should be given but also the buffers in the electrode chambers. In the case of aqueous solutions, it is advisable to specify the water used, e.g., double distilled water, freshly boiled to remove dissolved CO_2 from the air, etc. [1]

Methods need to provide detailed, step-by-step protocols for preparing, storing, and validating the mobile phases.

The mobile phase in CEC appears to be even more complex than in HPCE. Until the basics are worked out, it will probably be necessary to resort to reproducing recipes exactly. This should yield validatable methods with useful precision. One can anticipate that improved understanding of the effect of mobile phase components and concentration will be understood and then exploited.

Still, HPCE can offer some guidance, as in the example of buffer selection. In general, mobile phases need to be buffered, since the electroosmotic flow, as well as the form of ionizable analytes, is pH dependent. Without buffering, the migration time may vary to an unacceptable extent. Hjerten examined physical parameters on applied voltage and capillary length and mobile phase conductivity in HPCE [2]. This led to the recognition that buffer agents made from zwitterions or ampholytes produce useful buffer capacity with low conductivity.

Table 1. Ampholytes Suitable as Buffering Agents. The pH of the Solution Is the pI of the Ampholyte

	pI	pK_1	pK_2	pK_3	pK_4
N-Cyclohexyliminodiacetic acid	1.9	1.62	2.22	10.59	
N-(1-Carboxycyclohexyl)-imino-diacetic acid	2.1	1.63	2.59	11.24	
Aspartyl-aspartic acid	3.0	2.70	2.40	4.70	8.26
Glycyl-aspartic acid	3.6	2.81	4.45	8.60	
Glycyl-L-histidine	7.5	2.66	6.77	8.24	
Glycyl-glycyl-L-histidine	7.5	2.84	6.87	8.22	
Hydroxy-lysine	9.1	2.13	8.62	9.67	
Lysine	9.7	2.18	8.62	9.67	

From Ref. [3], with permission.

This enables the use of high field strength (V/cm). The net result is a reduction in run time of more than 80%.

Some compounds that provide low-conductivity buffers are presented in Table 1. In general, the buffer concentration should be as low as practical, usually 1 to 10 mM. This is illustrated in Figure 1 [3].

3.1 Bubble Prevention

Bubble formation can be a practical problem in CEC. Initially, bubbles formed sporadically and spontaneously. This led to instruments that pressurized both ends of the capillary to about 10 bar. As with HPLC, this pressurization prevented the formation of bubbles. Pressurization of the outlet end was a problem with interfacing to a mass spectrometer. A report by Poppe et al. of the University of Amsterdam showed that frits were the source of most of the bubbles [4]. They found that frits

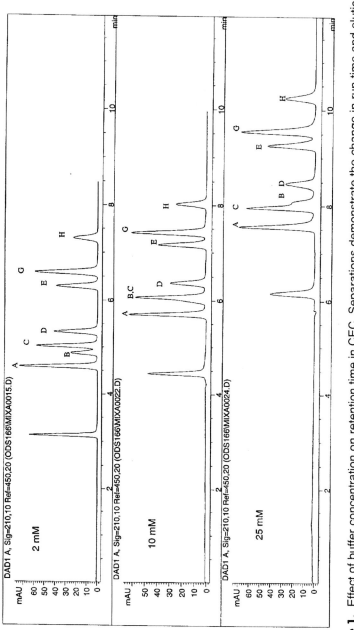

Figure 1. Effect of buffer concentration on retention time in CEC. Separations demonstrate the change in run time and elution order as the mobile phase electrolyte concentration increases. Column: Spherisorb ODS-1, 3 μm (75 μm × 25 cm, 34.5 cm total). Mobile phase: 60:40 ACN: (a) 2 mM KH_2PO_4 (pH 3.0), (b) 10 mM KH_2PO_4 (pH 3.0), (c) 25 mM KH_2PO_4 (pH 3.0). Conditions: 30 kV, 10 bar pressure, 35°C. Detection: UV at 210 nm. Sample: thiourea and compounds at mg/mL, injected 2 kV for 12 s. (From Ref. [3], with permission.)

constructed from the same material as packed in the capillary reduced bubble formation. Perhaps the frit material produced a different electroosmotic flow, which caused cavitation. Poppes group also found that sparging the mobile phase with helium reduced problems with bubbles. This trick is also used in HPLC to keep bubbles from forming in detector cells.

3.2   Effect of Organic Solvents

In HPLC, elution is controlled by adjusting the concentration of organic solvents. In the RPLC mode, increasing the organic solvent concentration increases the solvating power of the mobile phase for less polar compounds. Increasing the organic content of aqueous eluents increases speed in accordance with the like-dissolves-like principle.

Similar effects are observed with CEC in the reversed-phase mode with aqueous eluents, as shown in Figure 2. Increasing the percentage of acetonitrile (ACN) from 60% to 75% shortens the run time by 50% but at the expense of resolution. Note that peaks B and C change elution order. One should take care in extrapolating these results to nearly pure organic eluents, since inversion of the effect has been observed.

In RPLC, ACN is the preferred modifier for the mobile phase, since the viscosity and UV absorbance are lower and ACN seems to provide a wider operating range of elution strength. In CEC, ACN also appears to be the preferred mobile phase modifier, as shown in Figure 3. While ACN offers faster elution, methanol offers better separation of drugs E, F, G, and H. It should also be noted that two of the drug substances (B and D) do not elute when methanol is used as the modifier. This is peculiar, since they do elute when the same column is used in HPLC. The reason for this behavior is not known.

The stationary phase and mobile phase are coupled in a complex manner. Until this process is better understood, the mobile phase in CEC will probably be the source of most reproducibility problems. In the absence of a thorough understanding of the mobile phase, one should fall back on carefully documented, detailed methods describing

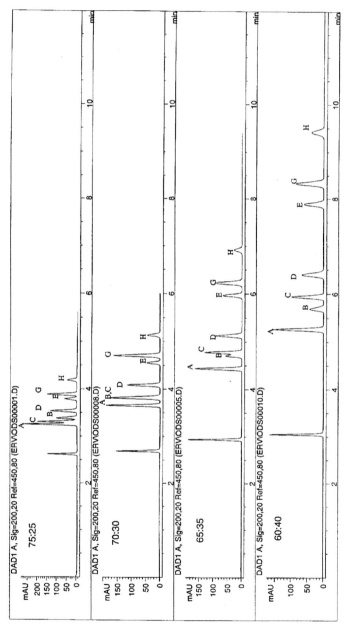

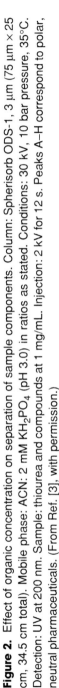

Figure 2. Effect of organic concentration on separation of sample components. Column: Spherisorb ODS-1, 3 μm (75 μm × 25 cm, 34.5 cm total). Mobile phase: ACN: 2 mM KH₂PO₄ (pH 3.0) in ratios as stated. Conditions: 30 kV, 10 bar pressure, 35°C. Detection: UV at 200 nm. Sample: thiourea and compounds at 1 mg/mL. Injection: 2 kV for 12 s. Peaks A–H correspond to polar, neutral pharmaceuticals. (From Ref. [3], with permission.)

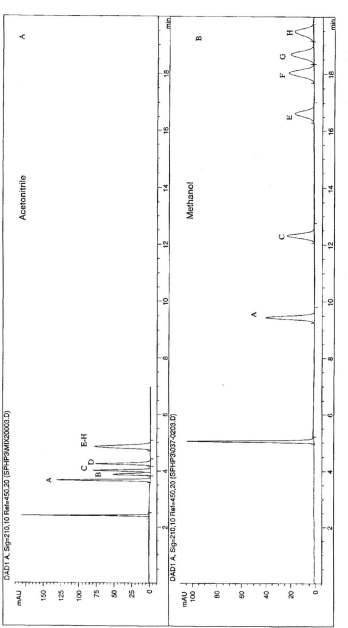

Figure 3. Comparison of CEC with aqueous methanol and acetonitrile. The separation demonstrates the resolving power of methanol on a phenyl phase. Column: Hypersil phenyl, 3 µm (75 µm × 25 cm, 34.5 cm total). Mobile phase: 60% (A) ACN, (B) methanol, 40% 2 mM KH_2PO_4 (pH 3.0). Conditions: 30 kV, 10 bar pressure, 35°C. Detection: UV at 210 nm. Sample: thiourea and compounds (1 mg/mL). Injection: 2 kV for 12 s. Peaks as in Figure 1. Note that peaks B and D do not elute in CEC but do elute in HPLC. (From Ref. [3], with permission.)

preparation and validation of the mobile phase. If the buffers are to be made by others, it is important that the operator variability be investigated. This should help reduce interlaboratory variance.

References

1. J. Beckers and P. Bocek, Letter to the Editor, *Electrophoresis, 19,* 2262 (1998).
2. S. Hjerten, L. Valtcheva, K. Elenbring, and J.-L. Liao, *Electrophoresis, 16,* 584 (1995).
3. P. D. Angus, E. Victorino, K. M. Payne, C. W. Demarest, T. Catalano, and J. F. Stobaugh. *Electrophoresis 19,* 2073 (1998).
4. S. E. van den Bosch, S. Heemstra, J. C. Kraak, and H. Poppe, *Journal of. Chromatography A, 755,* 165 (1996).

4 Instrumentation

From the early days when Pretorius first demonstrated the concept of electrochromatography using 1-mm-i.d. packed columns, there have been significant developments in columns and instruments for CEC [1]. Improvements in column manufacture have led to capillary columns with dimensions that minimize Joule heating. Instruments have been improved as well. The all too familiar problem of bubble formation at high electric field strength, which has been attributed to degassing of the mobile phase due to heating and differences in electroosmotic velocities through the packed bed, frit, and open section [2–5], has been suppressed with the modification of the instrument by pressurization of the buffer vials and improvements in column packing methods.

A variety of approaches have been explored for performing CEC separations, ranging from the most common isocratic elution methods to step gradients and true gradient elution. Pressurized flow capillary electrochromatography has also been employed, along with the various detection schemes. This chapter covers the instrumentation for separation and detection by various modes of CEC. Examples of applications using these instruments are presented in Chapters 5 and 6.

4.1 General Instrumentation

4.1.1 Isocratic CEC

The instrumentation for CEC in the isocratic elution mode is relatively simple. In fact, CEC can be performed using existing capillary electrophoresis instruments. The instrumentation consists of basically four parts, as shown in Figure 1: a capillary for performing the separation, a high-voltage power supply to generate the electroosmotic drive, a detector, and some form of a safety mechanism to protect against high voltage [6–7]. Numerous texts have been published regarding the general instrumental arrangement for CE [7–10].

For the most part, the instruments used today for performing CEC are still laboratory built. This is simply a reflection of the fact that CEC is still in its academic stages. The few commercial instruments marketed for CEC (notably by Hewlett-Packard and Beckman) are basically CE instruments that have been modified to allow for

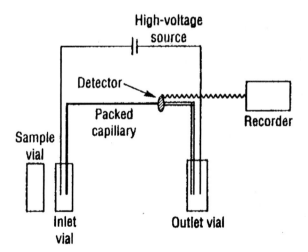

Figure 1. Schematic diagram of a CEC system. (From Ref. [7], with permission.)

pressurization of the inlet and outlet vials [2,11]. It should be noted that the pressure is applied to prevent bubble formation; it is not the driving force for flow of the mobile phase, as in PEC. Pressures of up to 12 bar may be applied to both ends of the capillary, resulting in no net pressure drop across the capillary.

Many groups have reported the use of conventional CE instrumentation, without modifications (e.g., pressurization), with varying degrees of success for performing CEC [12–15]. In these cases, care is taken to thoroughly degas the mobile phase by using vacuum and ultrasonication or helium sparging prior to use in CEC. In addition, use of low-conductivity buffers minimizes Joule heating and bubble formation [13]. Temperature control is also important for obtaining reproducible results and for minimizing Joule heating. Most commercial CE instruments today have built-in temperature control mechanisms. The instruments, either unmodified or with pressurization capabilities, allow for simple isocratic or step-gradient elution. Conditioning of the column often requires the capillary to be removed from the instrument and flushed with the mobile phase using an HPLC or syringe pump. Conditioning of the columns by applying a potential is also performed and is critical to successful CEC with either a commercial or laboratory-built system.

Injection of the sample is usually achieved by electromigration methods since hydrostatic and vacuum methods do not work well in CEC because of the high back pressure produced by packed columns. Hydrodynamic injections, by applied pressure, can also be employed if the instrument capabilities are available [16].

4.1.2 Step-Gradient CEC

To expand the applicability of CEC for samples of complex mixtures containing analytes of widely different hydrophobicities, gradient elution capability is desirable. The simplest approach is to set up a step gradient with existing CE instruments. The capillary column is equilibrated with the initial mobile phase/electrolyte, and the sample is injected. To form the step gradient, the voltage is removed at a point in the analysis, the inlet and outlet vials are replaced by ones containing a different mobile phase, and the potential is reapplied for continued analysis. For sequential steps, this vial exchange process is repeated at given time intervals. At the end of the analysis, after the last peak has eluted, the initial vials are replaced and the capillary is reequilibrated for the next analysis. The various commercial instruments available allow for all of this to be automated by programming a time-table of events [2]. Figure 2 illustrates the repeatability of an electrochromatographic analysis performed using an automated step gradient on a

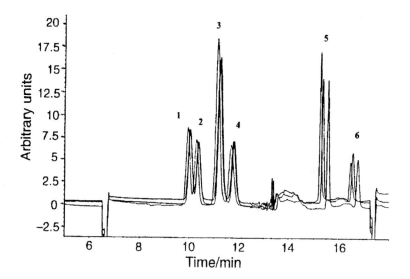

Figure 2. Repeatability of step-gradient CEC separation of six commonly used diuretic compounds. Conditions: 23.0 cm × 50 μm, CEC Hypersil C18, 3 μm; MeCN-50 mM Na_2HPO_4, pH 2.5-H_2O (Initial 40:20:40 for 0–6.50 min, 60:20:20 for 6.50–17.25 min, 40:20:40 for 17.25–25.00 min); 30 kV, 8 bar pressure both ends; 15°C. Detection: UV, 210 nm. Injection: 5 kV/15 s. Sample: 0.2 mg/mL, chlorothiazide (1), hydrochlorothiazide (2), chlorthalidone (3), hydroflumethiazide (4), bendroflumethiazide (5), bumetadine (6). (From Ref. [17], with permission.)

commercial instrument; the relative standard deviations for the retention times for repeat injections were below 1% [17].

Typical reversed-phase analyses start with a low-organic-strength mobile phase initially to elute the hydrophilic components, followed by switching to mobile phases of increased organic strength to elute the more hydrophobic analytes. The stopping of the flow by switching off

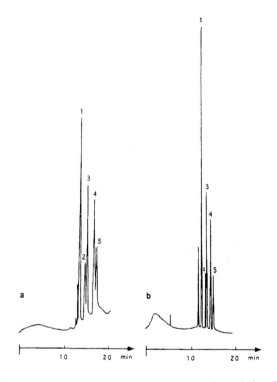

Figure 3. Comparison of isocratic (a) and step-gradient elution (b) CEC for separation of polycyclic aromatic hydrocarbons. (a) MeCN-4 mM sodium phosphate, pH 7.4 (60:40). (b) Initial: MeCN-4 mM sodium phosphate, pH 7.4 (50:50); after 6 min: MeCN-4 mM sodium phosphate, pH 7.4 (70:30). Conditions: 16.0 (20.0) cm × 75 μm, C18 derivatized continuous bed containing sulfonic acid groups; 3.0 kV. Detection: UV. Injection: 1 kV/2 s. Sample: (1) naphthalene, (2) 2-methylnaphthalene, (3) fluorene, (4) phenanthrene), (5) anthracene. (From Ref. [20], with permission.]

the voltage does not appear to affect the efficiency of the analysis [18–19]. Step-gradient elution has also been used to enhance the separation via zone sharpening, to increase the resolution, and to reduce the overall analysis time, as illustrated in Figure 3 [20]. The analysis is performed with the initial mobile phase of 50:50 acetonitrile/4 mM phosphate buffer, pH 7.4. After six minutes, the inlet vial was replaced by 70:30 acetonitrile/4 mM phosphate buffer, pH 7.4 (Figure 3b). By comparison, Figure 3a is an electrochromatogram of the same test mixture under isocratic conditions using 60:40 acetonitrile/4 mM phosphate buffer, pH 7.4. The sharpening of the starting zone, along with gradient elution, resulted in baseline separation.

4.1.3 Continuous-Gradient CEC

In order to perform a true, continuous-solvent gradient in CEC, a system can be constructed in-house by combining parts and supplies commercially available for HPLC and CE. There are two general approaches to forming a gradient in CEC. The first is that reported by the groups led by Bayer, Horvath, and Taylor [21–24]. In this method, a continuous gradient is formed with the use of HPLC pumps and is then sampled by the CEC capillary at the inlet reservoir. The composition of the mobile phase is thus varied in a manner similar to conventional HPLC. The tee is maintained at ground potential to prevent damage to the HPLC pumps. A length of fused silica tubing with narrow i.d. serves as the restrictor and waste line. A schematic of this approach is described in Figure 19 on p. 25. A slightly modified instrumental arrangement is illustrated in Figure 4. Here a stainless steel tee serves to split the flow from the HPLC pump and also as the inlet reservoir for the CEC capillary, as detailed in Figure 5. The tee is connected to a high-voltage power supply and placed in a plastic box fitted with a safety cut-off switch. The back pressure resulting from the flow-restriction capillary (10–15 bar) is adjusted by the split ratio and the HPLC pump flow rate to prevent bubble formation. This pressure is not sufficient to generate significant head pressure on the CEC capillary or a significant hydro-

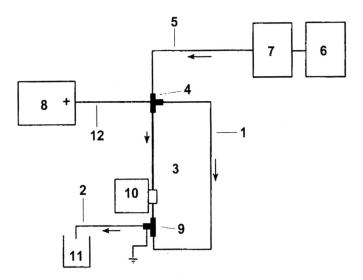

Figure 4. Continuous-gradient CEC system with a tee connection as the sampling interface. Arrows indicate the direction of flow. (1,2) Waste/restrictor capillary, (3) CEC separation capillary, (4) sampling interface, (5) loading capillary, (6) gradient HPLC pump, (7) HPLC autosampler, (8) high-voltage power supply, (9) grounded waste interface, (10) UV detector, (11) waste reservoir, (12) HV power cable. (From Ref. [22], with permission.)

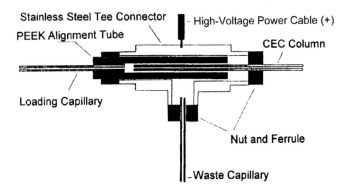

Figure 5. Cross-section of the CEC sampling interface. Sample and mobile phase are delivered to the tee, which is held at high voltage. The liquid is sampled electrokinetically by the CEC column, with the majority passing to the waste. (From Ref. [23], with permission.)

dynamic flow. Horvath's group has used a PEEK cross for similar purposes, as illustrated in Figure 6. Samples on these gradient systems are injected using the standard six-port rotary valve, and a fraction of the sample plug is introduced to the separation capillary. Thus, autosamplers can be utilized to improve injection reproducibility.

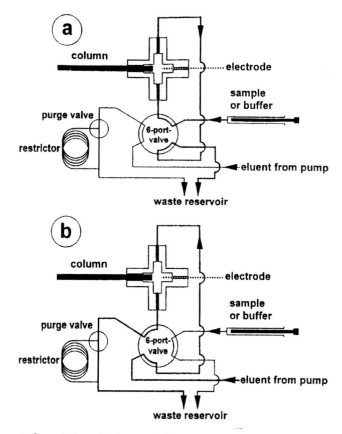

Figure 6. Sample introduction and eluent flow control system. (a) Arrangement for purging the inlet reservoir with sample solution for subsequent electrokinetic injection and for flushing the reservoir with the mobile phase. (b) Arrangement for operation in the electrochromatographic mode with isocratic or gradient elution. From Ref. [24], with permission.)

Dorsey et al. have described a very similar approach to that of Bayer, Horvath, and Taylor and have applied it to both aqueous and nonaqueous CEC applications and separations [64]. A flow injection analysis–CEC interface was used for gradient elution CEC, giving purely electroosmotic flow through the analytical column. Solvent gradients were performed with a micro-LC system connected to the interface. Peak tailing and peak width increased slightly compared with conventional instrumentation. More importantly, reproducibility of retention times in eight replicate injections was found to be better than 2% RSD for all solutes. This is somewhat better reproducibility than for most other gradient systems described in the literature.

Another approach, developed by Zare's group (see Figure 17, p. 23), the voltage applied to each of the mobile-phase reservoirs is varied in order to generate an electroosmotically driven solvent gradient [25]. The electroosmotic flows (EOFs) were generated in open fused-silica capillaries and regulated by computer-controlled voltages. A tee connector serves as the mixer. As the potential applied to one reservoir is increased, the other is decreased in order to form a smooth, continuous gradient. In this case, the final composition of the eluent is not readily obtained. However, this is not a major concern as long as the system is reproducible. Samples are introduced directly into the separation capillary electrokinetically by disconnecting it from the tee connector and placing the inlet into the sample vial. The advantage of this approach is that there is little loss of sample or wasted solvents since there is no splitting of the mobile phase; however, the injection process is not automated, a potential disadvantage.

The reproducibility of gradient formation using the HPLC pump approach is presented in Figure 7 [24]. The gradient-forming mobile-phase reservoir was spiked with 5% acetone, and the baseline was monitored by UV detection. The delay time can be readily minimized using low–dead volume connectors.

As the mobile phase is continuously swept by the capillary inlet, electrolyte depletion and the resulting reproducibility problems are no

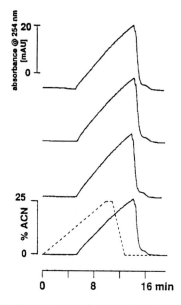

Figure 7. Reproducibility of the gradient profile. Conditions: 11 (14.4 total) cm × 50 μm, Zorbax ODS, 5 μm; solvent A: 10 mM phosphate, pH 7.0; solvent B: 10 mM phosphate, pH 7.0-acetonitrile-acetone (70:25:5); gradient: 0 to 100% B in 10 min, 100% B for 1 min, 100% to 0% B in 1 min at 0. 1 mL/min; 10 kV; 25°C. Detection: UV, 254 nm. (From Ref. [24], with permission.)

longer an issue, allowing for the use of low-ionic-strength mobile phases, which are desirable when interfaced with mass spectrometric detection. In addition, sample stacking of dilute samples at the head of the capillary prior to elution with increasing organic strength can be used to improve trace analysis [22].

4.1.4 PEC

Many of the problems encountered in CEC, such as bubble formation, analysis of anionic species, and long analysis time due to low EOFs for certain stationary phases, can be either minimized or eliminated if pressure is applied to the capillary column during the separation. In

PEC, the driving force for the mobile phase is a combination of pressure and EOF. The resulting effect of the combined flows on efficiency is demonstrated in Figure 8 [26]. The parabolic flow profile of the pressure component (LC) is superimposed on the plug-like flow profile of CE. Figure 9 illustrates the enhanced selectivity and improved efficiency offered by PEC vs. micro-HPLC, a purely pressure-driven system.

The instrumentation for PEC is essentially the same as that for gradient CEC [26–29] (Figure 10). Here again, an HPLC pump is used to push the mobile phase through the capillary columns while applying

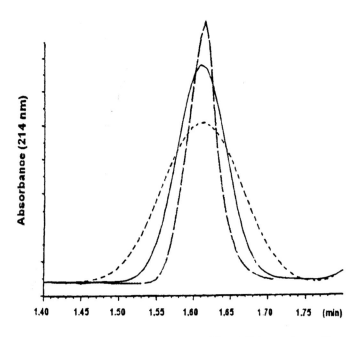

Figure 8. Elution profile of thiourea in capillary LC (- - - - -), capillary zone electrophoresis (CZE) (– – –), and PEC (———). Conditions: 15.0 (30.0) cm × 300 μm Octadecyl Si500 Polyol, 10 μm; MeOH-10 mM disodium tetraborate, pH 8.5 (75:25); for PEC: –8 kV and 25 bar pressure. To adjust the same mobile phase velocity in capillary LC: 34 bar. CZE: 12 kV. (From Ref. [26], with permission.)

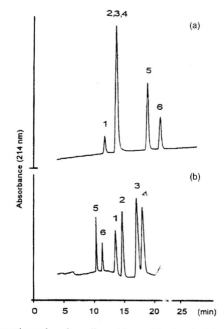

Figure 9. Separation of carboxylic acids and hydrophobic compounds with capillary LC (a) and PEC (b). Conditions: 15.0 cm × 100 μm Nucleosil 100 3-C8, 3 μm; MeOH-20 mM disodium tetraborate, pH 8.5 (75:25); 63 bar (a), –6 kV and 63 bar (b). Samples: 0.5–1 mg/mL, (1) folic acid, (2) p-hydroxy-benzoic acid, (3) acetylsalicylic acid, (4) nicotinic acid, (5) thiourea, (6) nicotinamide. (From Ref. [26], with permission.)

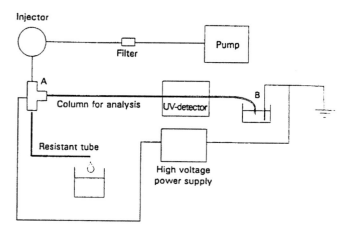

Figure 10. PEC apparatus. (From Ref. [28], with permission.)

a potential. This setup not only allows for gradient elution, it also simplifies operation since the columns can be flushed and conditioned on the instruments.

With the addition of pressurized flow, anionic species that tend to migrate against the EOF can be analyzed with improved analysis times. These components elute at long retention times, if at all, when using high-pH mobile phases, conditions conducive to high EOFs in CEC. Using low pHs, the anionic analytes can be separated and detected, though again, with longer analysis times due to the pH dependence of the EOF. In addition, PEC allows for hydrodynamic injection of the samples, which overcomes the major disadvantage of electrokinetic injections: discrimination of analytes based on charges. This should lead to more reproducible separations.

4.2 Capillary Columns

The heart of the CEC system is the column. (Refer to Chapter 3 for packing, frit formation, supports, and phases.) It is emphasized that the conditioning and proper flushing of the capillary are essential to successful CEC operation. Typically, the columns are flushed/conditioned with the run buffer/mobile phase under pressure using either HPLC or manual syringe pumps. The vacuum purge mechanism available with many commercial CE instruments cannot be used to accomplish this because of the high back pressures generated by the packed bed. Pressure may be used if the capabilities are available. The columns are further conditioned by application of the potential, slowly ramping the voltage from low to high and holding until a stable current is obtained. Alternatively, columns have been conditioned using just the electroosmotic flow [4, 12, 30–33].

The total length of the capillary column and the length to the packed bed (effective length) are determined by the instrumental configuration. Typically, the optical detection window is made immediately after the packed bed in order to minimize extra-column CE separations. For commercial instruments, there is a minimum length required to properly fit the capillary in the electrolyte reservoirs and align the detection window, often requiring long columns. The use of short-end injection techniques overcomes this limitation and allows for the use of shorter packed lengths. The injection is performed analogous to CE, at the outlet end, with the polarity of the applied voltage reversed [34]. With mass spectrometric detection, short columns can be readily employed since a lower resolution and peak capacity are better tolerated in this detection mode.

4.3 Detection

All detection schemes available for CE should potentially be applicable to CEC. Ultraviolet (UV), UV photodiode array (UV-PDA), and laser induced fluorescence (LIF) detection are commonly used, and CEC has been interfaced with MS and even NMR. Because of the narrow peak widths, fast sampling rates for data acquisition are required. Engelhardt et al. recommend a rate of 25 Hz or higher for fast separations [35].

4.3.1 UV and Fluorescence

Systems for performing UV and fluorescence detection are essentially the same as those used for CE [3, 5, 7, 25, 36]. For optical detection, windows are made on the capillary column by removing the polyimide coating. This is accomplished through a variety of methods, including burning off the coating by electric heating, chemical leaching with acid and gentle heat, and scraping off the coating with a blade [3, 24, 37–38]. Detection can be performed either on-column (after the packing material) or in-column such that the optical path is through the stationary phase [32, 37]. In-column detection eliminates the problem of peak broadening due to the detector cell and tailing due to interactions with the outlet frit. However, increased noise and decreased sensitivity and linearity may all result due to light scattering from the stationary phase particles [39–40]. The placement of the detection window can also be

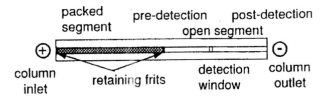

Figure 11. CEC column design. (From Ref. [41], with permission.)

used to add another dimension to the separation by allowing for CE separation in the open tube segment predetection (Figure 11). This aspect has been examined in detail through experimental and theoretical considerations [41–43].

The sensitivity of the UV detection system in CEC is limited by the optical path length (Beer's law), as in CE. The diameter of the column used for the separation cannot be significantly increased, as effective heat dissipation is required to minimize Joule heating. High-sensitivity Z- and bubble cells have been employed to bring sensitivity of CEC into the realm of HPLC [2, 44].

4.3.2 MS

Several groups have reported the interfacing of CEC and PEC with various modes of MS [18, 23, 45–57]. CEC-MS is an area in which the full potential of CEC may be truly realized. The coupling of CEC to MS provides a viable alternative to micellar electrokinetic chromatography for separation and detection of neutral analytes. The surfactants used for micelle formation in MEKC are incompatible with MS ion sources and complicate the analysis.

One of the interface designs for electrospray ionization is presented in Figure 12 [50]. A standard CE/MS triaxial probe fitted with an electrospray source was used. The probe consists of two concentric stainless steel capillaries surrounding the CEC separation capillary. The inner stainless steel capillary delivers the make-up sheath flow,

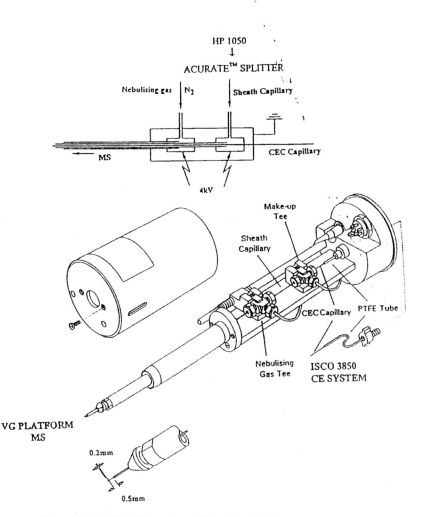

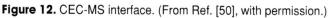

Figure 12. CEC-MS interface. (From Ref. [50], with permission.)

while the outer capillary provides the nebulizing gas. A modified version of this interface that allows for the use of short columns has also been described [51]. However, the sheath flow design leads to dilution of the analyte peak and is therefore not ideal. Sheathless systems have also been developed and applied to CEC, as described in Figure 13 [52].

Lubman et al. have successfully coupled gradient PEC with ion trap storage-reflectron time of flight mass spectrometer [53–56]. Figure 14 is a total ion chromatogram of a tryptic digest of chicken ovalbumin,

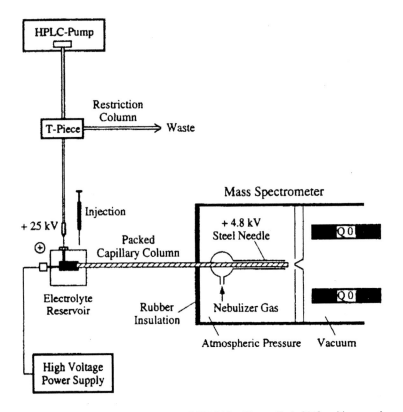

Figure 13. Setup for sheathless CEC-MS. (From Ref. [52], with permission.)

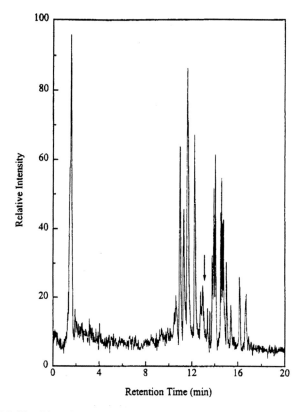

Figure 14. Total ion chromatogram of PEC separation of a tryptic digest of chicken ovalbumin with a sample injection amount of 12 pmol corresponding to the original protein. Conditions: 6 cm × 180 μm Vydac C18, 3 μm; solvent A: 0.07% TFA in H_2O, solvent B: 0.07% TFA in MeCN, 0 to 40% B in 20 min; 1 kV and 40 bar. (From Ref. [54], with permission.)

and Figure 15 presents the corresponding mass spectra for the peak indicated.

4.3.3 Nuclear Magnetic Resonance

The coupling of CEC with nuclear magnetic resonance (NMR) has also been demonstrated. Figure 16 illustrates the general instrumentation in which a special capillary containing an enlarged detection cell is used

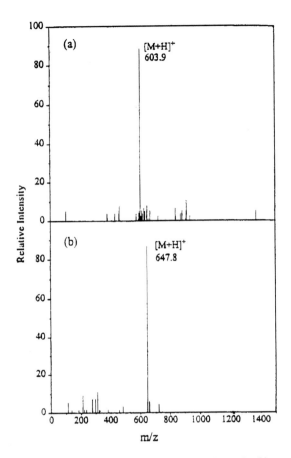

Figure 15. Mass spectra corresponding to the peak marked by an arrow in Figure 14. (a) Fragment TQINK, m/z = 603.9, eluted at 12.95 min. (b) Fragment VYLPR, m/z = 647.8, eluted at 13.01 min. (From Ref. [54], with permission.)

for on-line NMR detection [58–59]. A magnified view of the detection cell arrangement is presented in Figure 17. The versatile system allows for capillary electrophoresis, micro-LC, and CEC to be performed with minor changes in the setup. Both one- and two-dimensional NMR spectra can be acquired using stopped-flow methods.

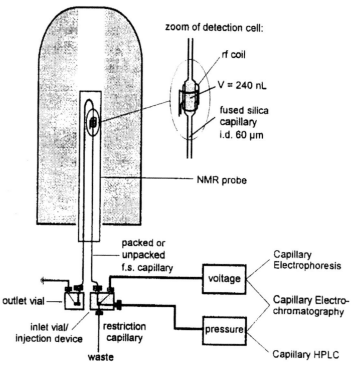

Figure 16. Instrumentation for coupling of capillary separation techniques to NMR. (From Ref. [58], with permission.)

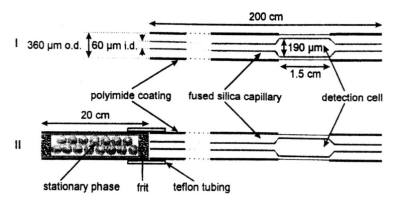

Figure 17. Capillary with enlarged detection cell for NMR detection. (I) The capillary is suitable for CE separations. (II) By connecting a packed capillary in front of the NMR detection capillary, the system is adapted to use with CEC and micro-HPLC. (From Ref. [58], with permission.)

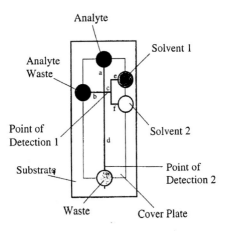

Figure 18. Microchip with channels and reservoirs indicated. The effective length of the main channel from the cross to "point of detection 2" is approximately 25 mm. (From Ref. [61], with permission.)

4.4 Microchips

An alternative approach to electrochromatography is a chip-based system. The design and operation of a microchip is detailed in the works of Ramsey et al. [60–62]. Essentially, narrow channels (3–10 mm) are formed in glass microchips, and the surface of these channels is modified with the stationary phase. The chips also contain sample, mobile phase, and waste reservoirs (Figure 18). Electroosmotic flow is used to load the sample and drive the mobile phase. Open-channel electrochromatography, isocratic and gradient, has been performed using the microchips, and Figure 19 [62] illustrates the results obtained for a test mixture of coumarin dyes using laser-induced fluorescence detection. Regnier et al. have also reported a novel column architecture for performing CEC on microchips [63].

4.5 Future Developments

In order for CEC to become a routine method for analysis, instruments specifically designed for CEC need to be commercially available. These

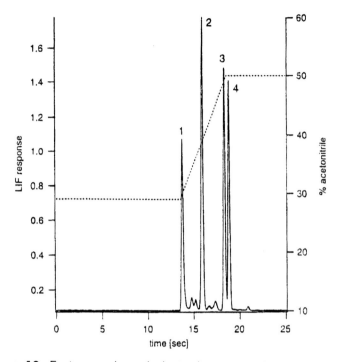

Figure 19. Fast open-channel electrochromatography on microchips. Channel depth, 5.2 μm; stationary phase, ODS; mobile phase, 10 mM borate, pH 8.4, with linear gradient of acetonitrile from 29 to 50% in 5 s, starting 1 s after injection; field strength, 700 V/cm. Samples: coumarin dyes, (1) C440, (2) C450, (3) C460, (4) C480. (From Ref. [62], with permission.)

instruments need to accommodate shorter column lengths typically used in CEC versus the longer columns standard for CE instruments. The ideal instrument would allow for the performing of isocratic and gradient CLC, CEC, and PEC, along with CE, all on a single platform. Some of the laboratory-built systems described in the literature are capable of combinations of CLC/CEC or CEC/PEC or CE/CEC. Another potential of CEC is the possibility of two-dimensional separations. By varying the length of the open (unpacked) segment of the capillary, it may be possible to perform a combination of CEC/CE in a single column.

References

1. V. Pretorius, B. J. Hopkins, and J. D. Schiek, *Journal of Chromatography, 99, 23* (1974).
2. http://www.agilent.com provides a good starting point for HPCE instrumentation and operations, and CEC as well, with application notes and references.
3. M. M. Dittmann and G. P. Rozing, *Journal of Chromatography A, 744, 63* (1996).
4. H. Rebscher and U. Pyell, *Chromatographia, 42,* 171 (1996).
5. J. H. Knox and I. H. Grant, *Chromatographia, 32,* 317 (1991).
6. J. H. Miyawa and M. S. Alasandro, *LC-GC Magazine, 16,* 36 (1998).
7. N. J. Dovichi, in *CE Theory and Practice,* P. Camilleri, Editor, CRC Press, Inc., Boca Raton, FL, 1993.
8. R. Weinberger, *Practical Capillary Electrophoresis,* Academic Press, Inc., San Diego, CA, 1993.
9. K. D. Altria, in *Capillary Electrophoresis Guidebook,* K. D. Altria, Editor, Humana Press, Inc., Totowa, NJ, 1996.
10. D. R. Baker, *Capillary Electrophoresis,* John Wiley & Sons, Inc., New York, NY, 1995.
11. Beckman technical literature on HPCE P/ACE MDQ capillary electrophoresis instrumentation, Beckman Coulter Instruments, Fullerton, CA, 1998–99, website: http://www.beckman.com/beckman/biorsrch/prodinfo/capelec/beckp12.asp.
12. L. A. Frame, M. L. Robinson, and W. J. Lough, *Journal of Chromatography A, 798,* 243 (1998).
13. S. E. Van der Bosch, S. Heemstra, J. C. Kraak, and H. Poppe, *Journal of Chromatography A, 755,* 165 (1996).
14. T. M. Zimina, R. M. Smith, and P. Myers, *Journal of Chromatography A, 758,* 191 (1997).
15. D. A. Stead, R. G. Reid, and R. B. Taylor, *Journal of Chromatography A, 798,* 259 (1998).
16. P. Sandra, A. Dermaux, V. Ferraz, M. M. Dittmann, and G. Rozing. *J. Micro. Sep., 9,* 409 (1997).
17. M. R. Euerby, D. Gilligan, C. M. Johnson, and K. D. Bartle, *The Analyst, 122,* 1087 (1997).
18. J. Ding and P. Vouros, *Analytical Chemistry, 69,* 379 (1997).

19. J. Ding, J. Szeliga, A. Dipple, and P. Vouros, *Journal of Chromatography A, 768,* 327 (1997).
20. J. L. Liao, L. Chen, C. Ericson, and S. Hjerten, *Analytical Chemistry, 68,* 3468 (1996).
21. B. Behnke and E. Bayer, *Journal of Chromatography A,* 680, 93 (1994).
22. M. R. Taylor, P. Teale, and S. A. Westwood, *Analytical Chemistry, 69,* 2554 (1997).
23. M. R. Taylor and P. Teale, *Journal of Chromatography A, 768,* 89 (1997).
24. C. G. Huber, G. Choudhary, and C. Horvath, *Analytical Chemistry, 69,* 4429 (1997).
25. C. Yan, R. Dadoo, and R. N. Zare, *Analytical Chemistry, 68,* 2726 (1996).
26. T. Eimer, K. K. Unger, T. and Tsuda, *Fresenius' Journal of Analytical Chemistry, 352,* 649 (1995).
27. S. Kitagawa, A. Tsuji, H. Watanabe, M. Nakashima, and T. Tsuda. *Journal of Microcolumn Separations, 9,* 347 (1997).
28. S. Kitagawa and T. Tsuda, *Journal of Microcolumn Separations, 6,* 91 (1994).
29. B. Behnke, E. Grom, and E. Bayer, *Journal of Chromatography A, 716,* 207 (1995).
30. M. R. Euerby, D. Gilligan, C. M. Johnson, S. C. P. Roulin, P. Myers, and K. D. Bartle, *Journal of Microcolumn Separations, 9,* 373 (1997).
31. M. T. Dulay, C. Yan, D. J. Rakestraw, and R. N. Zare, *Journal of Chromatography A, 725,* 361 (1996).
32. C. Yang and Z. El Rassi, *Electrophoresis, 19,* 2061 (1998).
33. N. W. Smith and M. B. Evans, *Chromatographia, 41,* 197 (1995).
34. M. R. Euerby, C. M. Johnson, M. Cikalo, and K. D. Bartle, *Chromatographia, 47,* 135 (1998).
35. H. Engelhardt, S. Lamotte, and F. L. Hafner, *American Laboratory, 40,* (1998).
36. M. M. Robson, M. G. Cikalo, P. Myers, M. R. Euerby, and K. D. Bartle, *Journal of Microcolumn Separations, 9,* 357 (1997).
37. M. M. Robson, S. Roulin, S. M. Shariff, M. W. Raynor, K. D. Bartle, A. A. Clifford, P. Myers, M. R. Euerby, and C. M. Johnson, *Chromatographia, 43,* 313 (1996).

38. R. J. Boughtflower, T. Underwood, and C. J. Paterson, *Chromatographia, 40,* 329 (1995).
39. H. Rebscher and U. Pyell, *Chromatographia, 38,* 737 (1994).
40. H. Rebscher and U. Pyell, *Journal of Chromatography A, 737,* 171 (1996).
41. G. Choudhary and C. Horvath, *Journal of Chromatography A, 781,* 161 (1997).
42. A. S. Rathore and C. Horvath, *Analytical Chemistry,* 70, 3069 (1998).
43. A. S. Rathore and C. Horvath, *Analytical Chemistry, 70,* 3271 (1998).
44. I. S. Lurie, R. P. Meyers, and T. S. Conver, *Analytical Chemistry, 70,* 3255 (1998).
45. S. E. G. Dekkers, U. R. Tjaden, and J. van der Greef, *Journal of Chromatography A, 712,* 201 (1995).
46. M. Hugener, S. P. Tinke, W. M. A. Niessen, U. R. Tjaden, and J. van der Greef, *Journal of Chromatography, 647,* 375 (1993).
47. D. B. Gordon, G. A. Lord, and D. S. Jones, *Rapid Communications in Mass Spectrometry, 8,* 544 (1994).
48. G. A. Lord, D. B. Gordon, P. Myers, and B. W. King, *Journal of Chromatography A, 768,* 9 (1997).
49. G. A. Lord, D. B. Gordon, L. W. Tetler, and C. M. Carr, *Journal of Chromatography A, 700,* 27 (1995).
50. S. J. Lane, R. Boughtflower, C. Paterson, and T. Underwood, *Rapid Communications in Mass Spectrometry, 9,* 1283 (1995).
51. S. J. Lane, R. Boughtflower, C. Paterson, and M. Morris, *Rapid Communications in Mass Spectrometry, 10,* 733 (1996).
52. K. Schmeer, B. Behnke, and E. Bayer, *Analytical Chemistry, 67,* 3656 (1995).
53. J. T. Wu, P. Huang, M. X. Li, M. G. Qian, and D. M. Lubman, *Analytical Chemistry, 69,* 320 (1997).
54. J. T. Wu, P. Huang, M. X. Li, and D. M. Lubman, *Analytical Chemistry, 69,* 2908 (1997).
55. P. Huang, J. T. Wu, and D. M. Lubman, *Analytical Chemistry, 70,* 3003 (1998).
56. J. T. Wu, M. G. Qian, M. X. Li, K. Zheng, P. Huang, and D. M. Lubman., *Journal of Chromatography A, 794,* 377 (1998).
57. G. Choudhary , C. Horvath, and F. J. Banks, *Journal of Chromatography A, 828,* 469 (1998).

58. K. Pusecker, J. Schewitz, P. Gfrorer, L. H. Tseng, K. Albert, and E. Bayer, *Analytical Chemistry, 70,* 3280 (1998).
59. K. Pusecker, J. Schewitz, P. Gfrorer, L. H. Tseng, K. Albert, E. Bayer, I. D. Wilson, N. J. Bailey, G. B. Scarfe, J. K. Nicholson, and J. C. Lindon, *Analytical Communications, 35,* 213 (1998).
60. S. C. Jacobsen, R. Hergenroder, L. B. Koutny, and J. M. Ramsey, *Analytical Chemistry, 66,* 2369 (1994).
61. J. P. Kutter, S. C. Jacobsen, and J. M. Ramsey, *Analytical Chemistry, 69,* 5165 (1997).
62. J. P. Kutter, S. C. Jacobsen, N. Matsubara, and J. M. Ramsey, *Analytical Chemistry, 70,* 3291 (1998).
63. B. He, N. Tait, and F. Regnier, *Analytical Chemistry, 70,* 3790 (1998).
64. A. S. Lister, C. A. Rimmer, and J. G. Dorsey, *Journal of Chromatography A, 828,* 105 (1998).

5 Applications for Small Molecules

The range of applicability and the applications with real samples will be the major factors in determining the success of CEC. In the past five years there has been a sharp increase in the number of papers demonstrating the applicability of CEC for various classes of compounds. This chapter summarizes the variety of approaches taken for CEC separations, depending on the nature of the analytes.

While there are references that deal with various classes of compounds [1–3], it should be noted that most of the literature to date deals with simply achieving the separation, primarily using standards. There are very few reports of actual quantitation or trace level determination using CEC for real samples. Also, as the technique of CEC is still in the research stages, the majority of the reports do not include data on various analytical figures of merit, besides plate counts and reproducibility of migration/retention times. Little has been described on selectivity values, limit of detection and quantification, or column-to-column reproducibility. And with the exception of a few reports, most CEC separations are performed using isocratic elution, since commercial CE instruments can easily be adapted for CEC operations. These instruments can also allow for simple step gradients. However, true gradient elution requires construction of a system in-house, or use of a commercial instrument with gradient capabilities for CEC.

5.1 Polycyclic Aromatic Hydrocarbons (PAHs) and Other Neutral Aromatic Compounds

The most common applications of CEC are separations of neutral analytes. Uncharged compounds are not separated by conventional capillary zone electrophoresis, and compared with HPLC, CEC offers the potential for higher efficiencies, better resolution, and shorter analysis times. Furthermore, the conditions required for generation of high EOF are often not ideal for separation of charged molecules. Acidic analytes tend to migrate against the EOF and require ion suppression for successful separation, and basic analytes interact with silanols on the capillary surface and/or packing, leading to peak tailing. A variety of approaches have been taken for separations using stationary phases, ranging from conventional reversed-phase packings to monoliths prepared in situ.

CEC has been very successfully applied to the separation of polyaromatic hydrocarbons (PAHs). In fact, PAHs are often used as test solutes to evaluate the columns performance in CEC (see Figure 1, p. 35). Columns packed with 3-μm C18 phases have been used to separate a mixture of parabens and PAHs, as illustrated in Figure 1 [4]. Baseline resolution was obtained for all of the components of the test mixture in under 7 minutes, and plate numbers ranged from 60,000 to 77,000. With laser-induced fluorescence (LIF) detection, Zare and colleagues studied PAHs and observed 400,000 plates/m with on-column detection [5]. The limits of detection were determined to be in the range of 10^{-9} to 10^{-11} M, with linear response (correlation coefficient of 0.9995) between 2×10^{-7} and 2×10^{-11} M for benzo[k]fluoranthene. In vitro reaction products of PAH deoxynucleoside adducts have also been separated using a C18 phase [68]. Both isocratic and step-gradient elutions were performed, and mass spectrometric detection was used for the identification of the adducts.

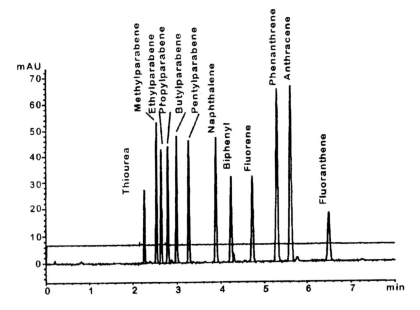

Figure 1. Isocratic separation of a model mixture containing 5 parabens, 6 PAHs, and thiourea as a marker. Conditions: packed length 25 cm (33.5 total length) × 100 mm i.d., CEC Hypersil C18, 2.5-µm particles; MeCN/25 mM MES, pH 6.0 (80:20 v/v); 20 kV applied voltage, 10 bar pressure applied to both ends of capillary; 20°C. Detection: UV, 250 nm. Injection: EK. Linear velocity: 1.9 mm/s. (From Ref. [4], with permission.)

Nonporous silica (NPS) octadecylsilane- (ODS) modified particles with small diameters (1.5 µm) have also been used for fast separations of 16 PAHs, classified as priority pollutants by the Environmental Protection Agency, as shown in Figure 2 [9–10]. Figure 3 illustrates the rapid separations possible using short columns [10]. By applying high voltages (55 kV), the same 16 PAHs were separated in under 3 minutes. It should be noted that problems with arcing are encountered when such high potentials are applied, justifying the use of low-conductivity buffers. Also, with such fast separations, data acquisition rates need to be adjusted upwards. This can require shorter time constants and higher frequency of data acquisitions. The problem of low sample loading and

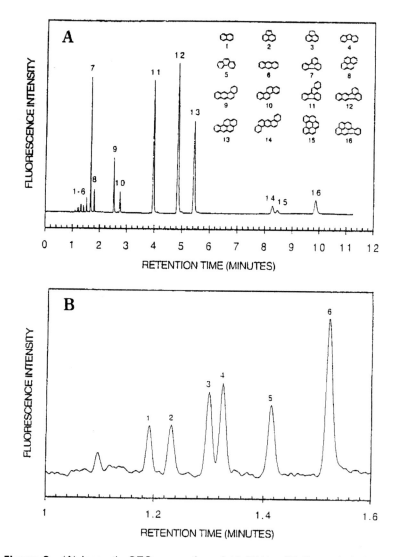

Figure 2. (A) Isocratic CEC separation of 16 PAHs. (B) Expanded view. Conditions: 20 (30) cm × 100 mm, nonporous ODS, 1.5 μm; MeCN/2 mM Tris, pH 9 (65:35); 29 kV. Detection: LIF. Injection: EK. (From Ref. [10], with permission.)

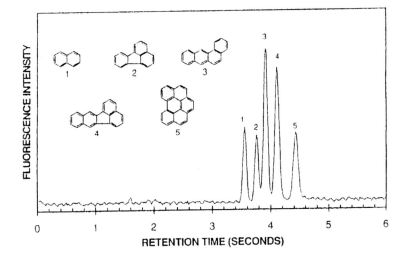

Figure 3. Electrochromatogram showing the rapid isocratic CEC separation of PAHs. Conditions: 6.5 (10) cm × 100 µm, nonporous ODS, 1.5 µm; MeCN/2 mM Tris, pH 9 (70:30); 28 kV. Detection: LIF. Injection: EK. Linear velocity: ~20 mm/s. (From Ref. [10], with permission.)

detection sensitivity with nonporous particles in CEC was circumvented by using LIF detection. The use of very small particles is limited for HPLC separations, because of the high pressure drop [11]. Alternatively, in electroosmotically driven CEC separations, 0.5-µm C8 bonded particles have been used for the separation of alkylbenzenes [12]. Figure 4 highlights the separation of 14 nitroaromatic and nitramine explosives and their degradates, using methanol/MES buffer with a nonporous, 1.5-mm ODS phase [13].

Hjertens group used columns with continuous polymer beds prepared in situ, called rigid polymer monoliths [1415]. Charged groups such as dextran sulfate and sulfonic acids were incorporated to produce EOF, and hydrophobic ligands such as C18 and C4 provided the selectivity. Since the bed was attached covalently to the capillary wall, the problems associated with frit formation, stability, and clogging were

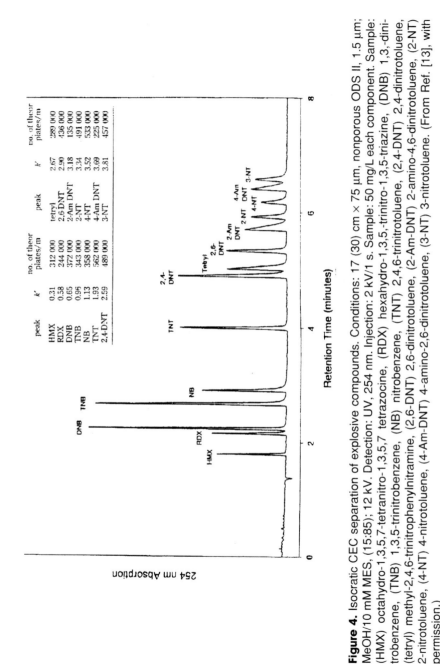

Figure 4. Isocratic CEC separation of explosive compounds. Conditions: 17 (30) cm × 75 µm, nonporous ODS II, 1.5 µm; MeOH/10 mM MES, (15:85); 12 kV. Injection: 2 kV/1 s. Sample: 50 mg/L each component. Sample: (HMX) octahydro-1,3,5,7-tetranitro-1,3,5,7 tetrazocine, (RDX) hexahydro-1,3,5,-trinitro-1,3,5-triazine, (DNB) 1,3,-dinitrobenzene, (TNB) 1,3,5-trinitrobenzene, (NB) nitrobenzene, (TNT) 2,4,6-trinitrotoluene, (2,4-DNT) 2,4-dinitrotoluene, (tetryl) methyl-2,4,6-trinitrophenylnitramine, (2,6-DNT) 2,6-dinitrotoluene, (2-Am-DNT) 2-amino-4,6-dinitrotoluene, (2-NT) 2-nitrotoluene, (4-NT) 4-nitrotoluene, (4-Am-DNT) 4-amino-2,6-dinitrotoluene, (3-NT) 3-nitrotoluene. (From Ref. [13], with permission.)

eliminated. Using a charged dextran sulfate polymer, efficiency of 120,000 plates/m was reported for the PAH separation. Addition of SDS to the acetonitrile/phosphate buffer, below the critical micelle concentration, enhanced the resolution and decreased the analysis time to under 10 minutes, as described in Figure 5. This may not be ideal for the case of MS detection. Gradient elution was also employed, resulting in zone sharpening and leading to baseline separation of all peaks. Frèchét et al. employed their own rigid polymer monoliths for the separation of benzene derivatives [1618].

Open tubular capillary electrochromatography (OT-CEC) has also been performed for the separation of PAHs [19]. The inner walls of the fused-silica capillary are coated with a ligand using the sol-gel method [2021]. The columns prepared with a C8 ligand were shown to be stable under acidic and basic conditions. Efficiencies ranging form 280,000 to 500,000 plates/m were obtained with a methanol/phosphate buffer.

Fluorinated columns using tridecafluoro-1,1,2,2-tetrahydrooctyl-1-triethoxysilane (F13-TEOS) have been prepared for use in the separation of halogenated organic compounds [22]. However, OT-CEC, in general, suffers from the problem of low sample loadability compared with the packed-bed approach because of the lower availability of surface area.

Maruska and Pyell demonstrated that CEC can also be performed in the normal-phase mode [23–24]. Using rigid octadecylated cellulose beads as the stationary phase and aqueous and nonaqueous eluents, they separated test mixtures consisting of PAHs and compounds with varying polarity. The efficiencies obtained using the cellulose-based phases were lower (N = 10,000 plates/m) than those obtained for silica phases (N = 160,000 plates/m) [25].

A variety of benzene derivatives and nitrogen-, oxygen-, and sulfur-containing heterocyclic analytes have also been examined by CEC [26–27]. The reproducibility of the retention times and the detector responses for the heterocyclic compounds are presented in Table 1. The values obtained were comparable to HPLC. The similarity in the

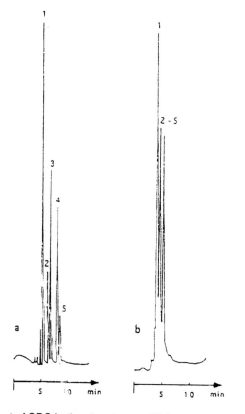

Figure 5. Effect of SDS in the eluent upon CEC separations of PAHs using monolithic polymer columns. (a) With addition of 1 mM SDS in the eluent. (b) Without SDS. Conditions: 30.0 (14.0) cm × 100 μm, C18 derivatized continuous bed containing sulfonic acid groups; MeCN/4 mM sodium phosphate, pH 7.4 (60:40); 3.0 kV. Detection: UV. Injection: 1 kV/2 s. Sample: (1) naphthalene, (2) 2-methylnaphthalene, (3) fluorene, (4) phenanthrene, (5) anthracene. (From Ref. [15], with permission.)

retention behavior of these compounds in CEC and HPLC modes opens the possibility of applying the rules of HPLC for method transfer to CEC.

The practical applicability of CEC for environmental pollutant monitoring has been explored, using insecticides and phthalate esters

Table 1. Reproducibility of Migration Times and Detector Responses

Compound no.	Compound name	% RSD	
		Migration time	Detector response
1	Quinolone	0.20	10.5
2	Indole	0.18	9.8
3	Isoquinolone	0.19	9.7
4	2-Methylquinoline	0.36	2.3
5	Carbazole	0.22	12.7
6	2,6-Dimethylquinoline	0.12	13.4
7	Acridine	0.20	10.1
8	Phenanthridine	0.34	5.9
9	7,8-Benzoquinoline	0.23	10.6
10	9-Methylcarbazole	0.25	12.3
11	Dibenzofuran	0.25	11.3
12	9-Ethylcarbazole	0.26	12.7
13	Dibenzothiophene	0.26	11.4

Number of determinations: 5. Compounds 4, 7, and 8 were in a separate mixture from the rest of the compounds in this table.

From Ref. [26], with permission.

as the test compounds on a variety of stationary phases [28]. In addition, column-to-column variability has been considered along with reproducibility of area and height or quantitation. For a given stationary phase, the column-to-column variability was high, as were the relative standard deviations for the injections. By using an internal standard, the precision could be significantly improved.

5.2   Pharmaceuticals and Related Analytes

5.2.1 Basic Compounds

For the majority of the reversed-phase packing materials used for CEC, EOF is due to the residual silanols in the packing and capillary walls. These silanols become problematic when separating basic compounds. Tremendous peak tailing occurs, and in some cases, the basic analyte may be adsorbed onto the stationary phase and may not elute within a reasonable amount of time. Lowering the pH results in some improvements in peak symmetry; however, the EOF is also reduced. For these reasons, a considerable amount of effort has been devoted to the CEC separation of ionic species.

Mobile-phase additives are used in HPLC separations to reduce tailing by competing with basic analytes for silanol interactions. These additives can also be used for CEC separations. Triethylamine (TEA) and triethanolamine (TEOA) have been effectively used with C18-packed columns and low-pH buffers for the separation of acidic, neutral, and basic compounds [29]. Figure 6 illustrates the dramatic difference produced when using TEA. Lurie and coworkers used hexylamine as the modifier; electrochromatograms of a simultaneous separation of acidic, basic, and neutral analytes with step-gradient elution are presented in Figure 7 [30]. The use of a bare silica phase has also been investigated for the separation of basic compounds, as shown in Figure 8 [31].

The ion exchange mechanism was exploited by Smith and Evans in their work on the separation of pharmaceutical compounds [32]. Various tricyclic antidepressants (Figure 9) and low-molecular-weight neutral and basic compounds were separated using a strong cation exchange phase (3-μm Spherisorb SCX). The sulfonic acid groups on the ion exchanger provided significant EOF, even at low pH, allowing for improved peak symmetry. Although not consistently reproducible, the report of reduced plate height of 0.04 and plate counts on the order of 8 million plates per meter clearly illustrated the potential of CEC. This was enough to generate significant interest in the area. Euerby et

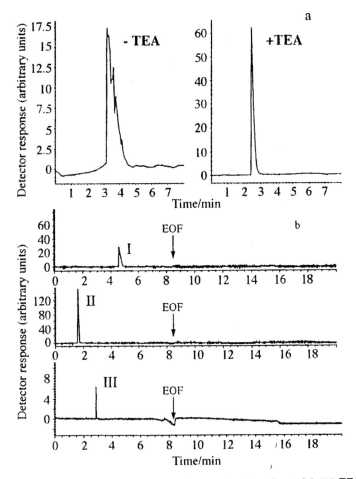

Figure 6. (a) Electrochromatograms illustrating the effect of 0.1% TEA in the eluent. Conditions: 25.0 (33.0) cm × 100 μm, CEC Hypersil C18, 3 μm; MeCN/50 mM KH<P7J10>2<P255J0>PO<P7J10>4<P255J0>, pH 2.3/H_2O (60:20:20) with 0.1% TEA; 20 kV, 8 bar pressure applied; 15?C. Detection: UV, 214 nm. Injection: 15 kV/5 s. Sample: 100 μg/mL, Astra Charnwood research compound. (b) Electrochromatograms of basic compounds. Conditions: 25.0 (33.0) cm × 100 μm, CEC Hypersil Phenyl, 3 μm; MeCN/50 mM KH_2PO_4, pH 2.3/H_2O (60:20:20) with 0.05% TEA; 25 kV, 8 bar pressure both ends; 15°C. Detection: UV, 214 nm. Injection: 15 kV/5 s. Sample: 1 μg/mL, nortriptyline (I), benzylamine (II), procainamide (III). Efficiencies: 6187, 29116, and 127,801 plates per column for I, II, and III, respectively. (From Ref. [29], with permission.)

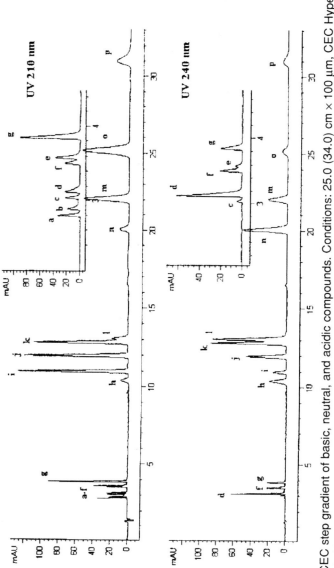
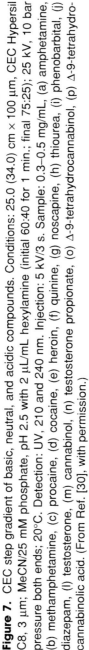

Figure 7. CEC step gradient of basic, neutral, and acidic compounds. Conditions: 25.0 (34.0) cm × 100 μm, CEC Hypersil C8, 3 μm; MeCN/25 mM phosphate, pH 2.5 with 2 μL/mL hexylamine (initial 60:40 for 1 min.; final 75:25); 25 kV, 10 bar pressure both ends; 20°C. Detection: UV, 210 and 240 nm. Injection: 5 kV/3 s. Sample: 0.3–0.5 mg/mL, (a) amphetamine, (b) methamphetamine, (c) procaine, (d) cocaine, (e) heroin, (f) quinine, (g) noscapine, (h) thiourea, (i) phenobarbital, (j) diazepam, (l) testosterone, (m) cannabinol, (n) testosterone propionate, (o) Δ-9-tetrahydrocannabinol, (p) Δ-9-tetrahydrocannabinolic acid. (From Ref. [30], with permission.)

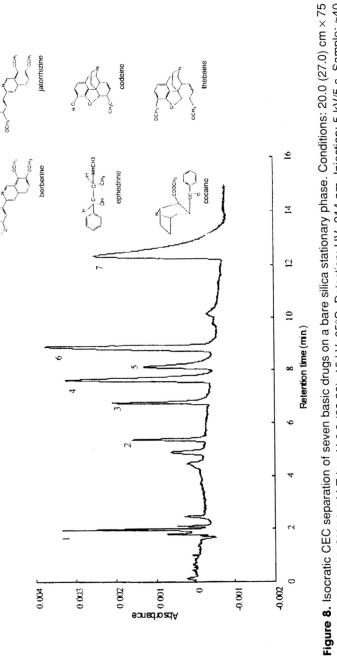

Figure 8. Isocratic CEC separation of seven basic drugs on a bare silica stationary phase. Conditions: 20.0 (27.0) cm × 75 µm, silica, 3 µm; MeCN/10 mM Tris, pH 8.3 (80:20); 15 kV; 25°C. Detection: UV, 214 nm. Injection: 5 kV/5 s. Sample: ~40 ppm each, (1) aniline, (2) cocaine HCl, (3) berberine HCl, (4) thebaine, (5) jatrorrhizine HCl, (6) ephedrine HCl, (7) codeine phosphate. (From Ref. [31], with permission.)

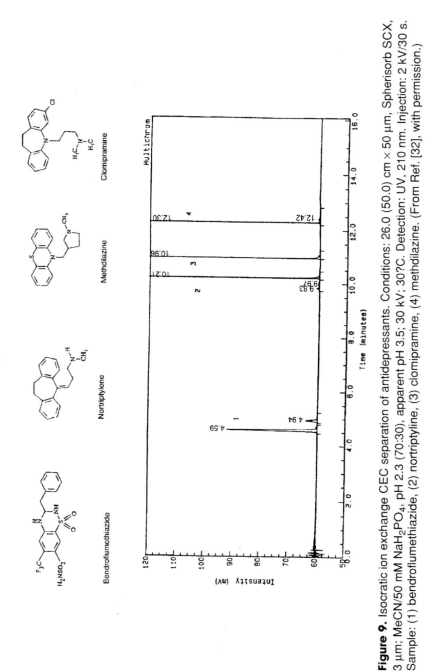

Figure 9. Isocratic ion exchange CEC separation of antidepressants. Conditions: 26.0 (50.0) cm × 50 μm, Spherisorb SCX, 3 μm; MeCN/50 mM NaH₂PO₄, pH 2.3 (70:30), apparent pH 3.5; 30 kV; 30°C. Detection: UV, 210 nm. Injection: 2 kV/30 s. Sample: (1) bendroflumethiazide, (2) nortriptyline, (3) clomipramine, (4) methdilazine. (From Ref. [32], with permission.)

al. have also reported efficiencies of over 16 million plates/m using the same stationary phase; however, the problem of reproducibility has not been solved [3]. Yan and coworkers, using the SCX phase, reported relative standard deviations for the retention times of less than 1% for the separation of various neutral and basic compounds [33]. Apparently, it is easier to obtain reproducibility of retention times, as opposed to very high efficiency and plate counts, in this particular mode of CEC.

5.2.2   Acidic Compounds

Separation of acidic compounds on conventional reversed-phase packings by CEC is complicated by the fact that at pHs required to ionize the silanols and generate strong EOFs (greater than 7), acids are anionic and hence migrate towards the anode. To overcome this, CEC analysis of acids has been performed in the ion-suppressed mode using low-pH (below the pKa of the analyte) electrolytes, as illustrated in Figure 7. Using this approach, the acids are uncharged and elute with the neutrals, and are separated due to differences in interactions with the stationary phase. Retention times, however, may be long since the EOF decreases with the decrease in pH for the reversed-phase electrochromatographic system.

Mixed-mode stationary phases containing both reversed-phase and ion exchange functionalities, such as C6/SCX or C18/SCX, overcome the strong pH dependence of the EOF, since the ion exchangers are ionized over a broad pH range. The mixed modes can thus be used for the separation of ionic species. Figure 10 compares the separations obtained using mixed-mode and reversed-phase systems [34]. Alternatively, pressure could be applied to push the analytes toward the detector, and PEC of carboxylic acids has been reported [35]. The pressure component of PEC, however, introduces a parabolic flow profile, as opposed to the flat profile obtained with CEC.

In terms of quantitative analysis, a CEC assay for paracetamol (acetaminophen) and aspirin tablets has been developed by Altria et al.

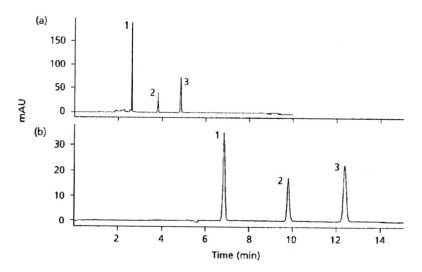

Figure 10. Comparison of isocratic CEC separation of acidic compounds using (a) mixed-mode and (b) reversed-phase columns. Conditions: 21.0 cm × 50 μm, (a) CEC Hypersil C18/SCX, 3 μm (b) CEC Hypersil C18,3 μm; MeCN/50 mM NaH_2PO_4, pH 2.3/H_2O (60:20:20); 30 kV, 8 bar pressure both ends; 15°C. Detection: UV-DAD. Injection: 5 kV/15 s. Sample: 0.3 mg/mL, (1) p-hydroxybenzoic acid, (2) bumetanide, (3) flurbiprofen. (From Ref. [34], with permission.)

[36]. A C18 phase with acetonitrile/pH 3.0 MES buffer was employed. Details are provided in Table 2 and Figure 11. An internal standard and high sample concentrations were used to improve quantitation. Unacceptable data were obtained when the results were calculated without the use of an internal standard. An impurity profile for the acidic diabetic drug troglitazone using CEC is illustrated in Figure 12 using a test mix spiked with impurities [36]. The CEC separation was performed isocratically, whereas the method used routinely for HPLC employs gradient elution. Levels of a specific impurity in a drug substance batch as determined by HPLC and CEC were comparable: 0.15% and 0.12% by area, respectively.

Table 2. Analytical Performance of an Assay Method for Aspirin and Paracetamol (Acetaminophen)

	Aspirin	Paracetamol
Precision RSD (%) (n = 10)		
Migration time (MT)	1.1	0.6
Relative MT	0.1	0.05
Peak area	3.0	1.6
Peak area ratio	0.3	0.34
Response factor	1.31	0.32
Linearity (50–150 ppm)	0.9995	0.9993
Label claim	299.3 mg/tablet	500 mg/tablet
Assay result	300 mg/tablet	497.5 mg/tablet

From Ref. [36], with permission.

5.2.3 Steroids

The application of CEC for the analysis of steroids has been investigated in detail by Wang et al. [37]. A method was developed and optimized for the separation of norgestimate and its related degradation products. The efficiencies for the main component and related impurities were in the range of 100,000 plates per meter. Linearity, limits of detection, and repeatability were evaluated. Correlation coefficients of 0.998 or better were obtained for each of the components, and the relative standard deviations for the retention time and the peak areas were less than 2%. Quantitation of 0.1% spiked impurities was achieved, as described in Figure 13, in half the analysis time of the existing HPLC method.

Successful separation has been achieved for the steroid tipredane and its related impurities on a Spherisorb ODS-1 packed column [3, 38]. In addition, CEC was able to baseline resolve tipredane from its C-17

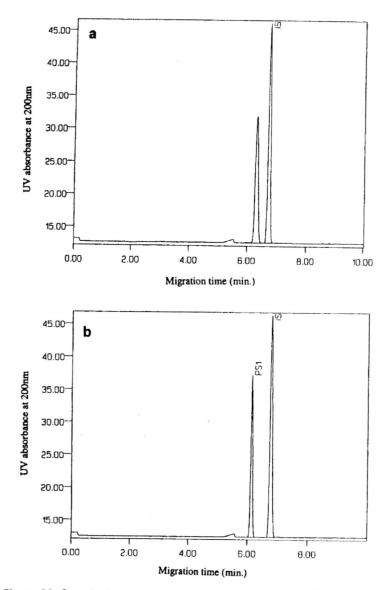

Figure 11. Quantitative separations of pharmaceuticals by CEC. Conditions: 20.0 cm × 50 μm, ODS; MeCN/10 mM MES, pH 3 (80:20); 15 kV. Detection: UV, 200 nm. Injection: 7 kV/5 s. Sample: (a) acetylsalicylic acid (aspirin), (b) paracetamol, IS = internal standard (benzamide). (From Ref. [36], with permission.)

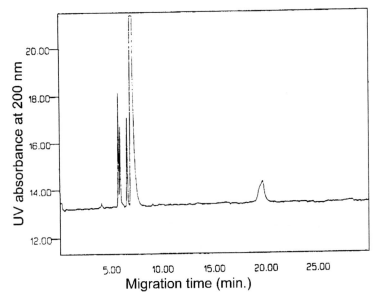

Figure 12. Isocratic CEC profile of an acidic pharmaceutical spiked with impurities. Conditions: 20.0 cm × 50 μm, ODS, 3 μm; MeCN/20 mM MES, pH 5.5 (75:25); 20 kV. Detection: UV, 200 nm. Injection: 7 kV/7 s. Sample: 3 mg/mL troglitazone. (From Ref. [36], with permission.)

diastereoisomer, with efficiencies of 174,000 and 181,000 plates/m, whereas HPLC using a variety of stationary phases failed to produce complete resolution.

A separation using both UV and MS detection of steroids, such as fluticasone propionate and related impurities, on Spherisorb ODS-1 and Hypersil ODS-Apacked columns has also been demonstrated [32, 3942]. Gradient-elution CEC of steroid hormones has also been described, as shown in Figure 14, using a Zorbax ODS packed column with an acetonitrile/pH 8.0 borate buffer [43].

ODS-modified 1.5-mm nonporous silica and 1.8-mm ODS-modified porous silica phases have been used for separations of test mixtures of steroids [4445]. As illustrated in Figure 15, the extremely fast separations were complete in about 3 minutes. The loading capacity of

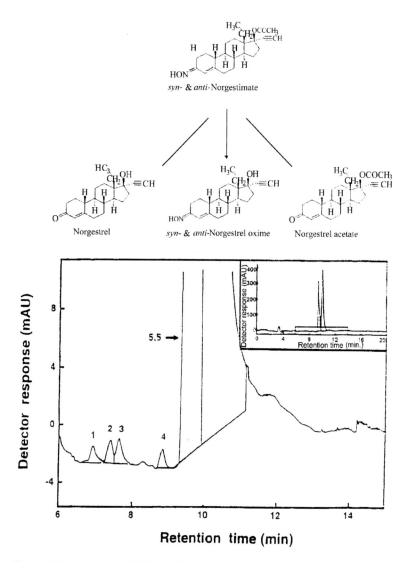

Figure 13. Isocratic CEC profile of norgestimate drug substance spiked with 0.1% of degradation impurities. Conditions: 25 (35) cm × 100 μm, C18, 3 μm; MeCN/THF/25 mM Tris HCl, pH 8/H_2O (35:20:20:25); 30 kV, 8 bar pressure both ends; 25°C. Detection: UV, 225 nm. Injection: 10 kV/3 s. Sample: (1) norgestrel, (2) *syn*-norgestrel oxime, (3) *anti*-norgestrel oxime, (4) norgestrel acetate, (5) *syn*-norgestimate, (6) *anti*-norgestimate. (From Ref. [37], with permission.)

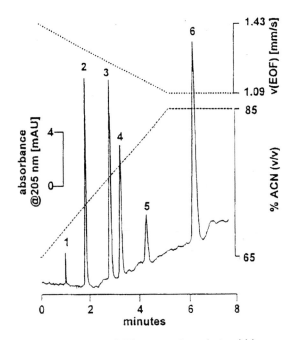

Figure 14. Gradient elution CEC separation of steroid hormones. Conditions: 9.6 (17.6) cm x 50 μm, Zorbax ODS, 6 μm; A: MeCN/10 mM borate, pH 8, (65:35), B: MeCN/10 mM borate, pH 8, (85:15), 0–100%B in 5 min, 100%B for 3 min at 0.1 mL/min; 14 kV; 25°C. Detection: UV, 205 nm. Injection: 1 kV/0.5 s. Sample: (1) ~1 mg/mL, formamide, (2) corticosterone, (3) testosterone, (4) androsten-3,17-dione, (5) androstan-3,17-dione, (6) pregnan-3,20-dione. Linear velocity: 1.43–1.09 mm/s. (From Ref. [43], with permission.)

the porous particles was about 50 times higher than that of the nonporous particles.

The application of CEC to biological sample matrices has been demonstrated using gradient elution to separate corticosteroids in urine and plasma [46]. Sample purification was essential to maintain column life. Figure 16 shows the separation and detection of hydrocortisone, dexamethasone, and fluocortolone in a sample of equine urine spiked with the steroids and extracted using solid-phase extraction. The linear-

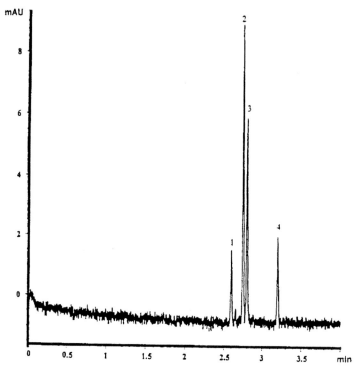

Figure 15. Isocratic CEC separation of a mixture of steroids. Conditions: 23.8 (32) cm × 100 μm, Chromspher ODS, 1.5 μm; MeCN/H$_2$O (60:40) + 5 mM SDS + 1. 6 mM sodium tetraborate, pH 9.25; 20 kV, 10 bar pressure both ends; 25°C. Detection: UV. Injection: 10 kV. Sample: (1) hydrocortisone, (2) testosterone, (3) 17-α-methyltestosterone, (4) progesterone. (From Ref. [44], with permission.)

ity for each component was established with a correlation coefficient of 0.998 or better, and the limit of detection was 0.39 μg/mL for all three components. CEC separations of samples of equine urine after administration of tetracosactrin acetate are presented in Figure 17 [46].

In another example using complex matrices, a reversed-phase LC method for the separation of progesterone and its metabolites in plasma has been successfully transferred to CEC [47]. A direct comparison of

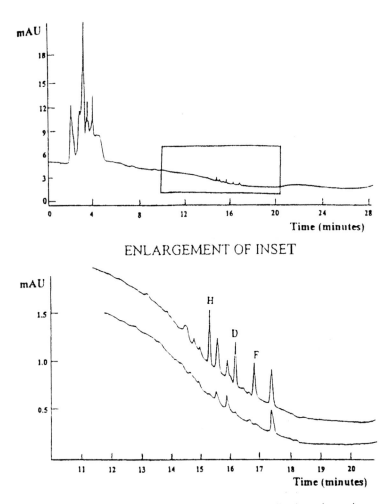

Figure 16. Gradient elution CEC separation of spiked equine urine extracted by SPE showing coextracted, weakly retained matrix peaks (2–5 min) and retention time region of corticosteroids (highlighted and expanded below). Upper electrochromatogram: spiked sample; lower electrochromatogram: blank urine extract. Conditions: 16 (24) cm × 50 μm, Apex ODS, 3 μm; A: H$_2$O, B: 5 mM ammonium acetate in MeCN, 1% to 80% B in 10 min at 0.1 mL/min; 25 kV; 20°C. Detection: UV, 240 nm. Injection: 50 μL. Sample: 0.4 μg/mL, (H) hydrocortisone, (D) dexamethasone, (F) fluocortolone. Linear velocity: ~1.5 mm/s. (From Ref. [46], with permission.)

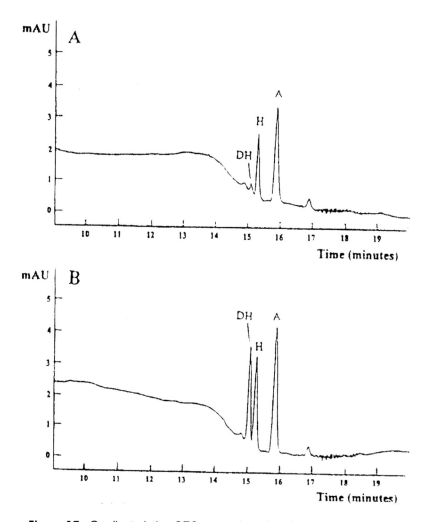

Figure 17. Gradient-elution CEC separation of equine urine samples taken after administration of tetracosactrin acetate and extracted by SPE. (A) 2 h postadministration, (B) 12 h post administration. Conditions: 16 (24) cm × 50 µm, Apex ODS, 3 µm; A: H$_2$O B: 5 mM ammonium acetate in MeCN, 9% to 80% B in 5 min at 0.1 mL/min; 25 kV; 20°C. Detection: UV, 240 nm. Injection: 100 µL. Sample: (DH) 10-β-dihydrocortisone, (H) hydrocortisone, (A) adrenosterone internal standard. (From Ref. [46], with permission.)

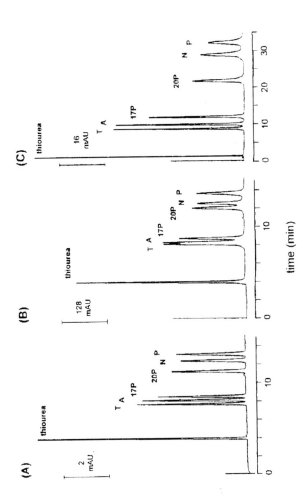

Figure 18. Comparison of (A) CEC and (B, C) HPLC separations of progesterone and its metabolites. (A) CEC conditions: 20 (35) cm × 100 μm, Hypersil ODS, 3 μm; MeCN/MeOH/20 mM Tris HCl, pH 8.0 (37.5:37.5:25); 15 kV. Detection: UV, 240 nm. Injection: 15 kV/5 sec (25 nL estimated). Linear velocity: 0.83 mm/s. (B) HPLC conditions: 20 cm × 4.6 mm, Hypersil ODS, 3 μm; MeCN/MeOH/20 mM Tris HCl, pH 8.0 (37.5:37.5:25); 0.6 mL/min. Detection: UV, 240 nm. Injection: 20 μL. Linear velocity: 0.83 mm/s. (C) HPLC using the same column as in B with the following change: MeCN-MeOH-20 mM Tris HCl, pH 8.0 (30:20:50); 2.1 mL/min. Samples: 20 μg/mL each, (P) progesterone, (17P) 17-α-hydroxyprogesterone, (20P) 20-α-hydroxyprogesterone, (A) androstenedione, (T) testosterone, (N) norethindrone internal standard, plus thiourea to marker. (From Ref. [47], with permission.)

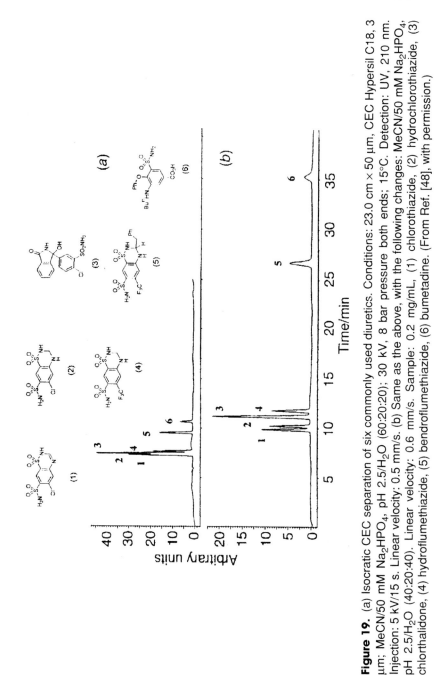

Figure 19. (a) Isocratic CEC separation of six commonly used diuretics. Conditions: 23.0 cm × 50 µm, CEC Hypersil C18, 3 µm; MeCN/50 mM Na₂HPO₄, pH 2.5/H₂O (60:20:20); 30 kV, 8 bar pressure both ends; 15°C. Detection: UV, 210 nm. Injection: 5 kV/15 s. Linear velocity: 0.5 mm/s. (b) Same as the above, with the following changes: MeCN/50 mM Na₂HPO₄, pH 2.5/H₂O (40:20:40). Linear velocity: 0.6 mm/s. Sample: 0.2 mg/mL, (1) chlorothiazide, (2) hydrochlorothiazide, (3) chlorthalidone, (4) hydroflumethiazide, (5) bendroflumethiazide, (6) bumetadine. (From Ref. [48], with permission.)

CEC and HPLC results was made (Figure 18). On-line preconcentration was achieved using low solvent strength (noneluting) solutions, in a manner analogous to HPLC.

5.2.4 Diuretics

CEC separations of mixtures of thiazide diuretics have been carried out using a C18-packed capillary (Figure 19) [48]. The pH of the acetonitrile/phosphate buffer was adjusted to 2.5 in order to elute the acidic analyte, which was negatively charged at the higher pH and hence migrated towards the anode. Step-gradient elution of the same mixture resulted in a shorter analysis time, as indicated in Figure 20.

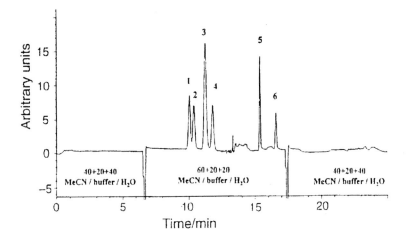

Figure 20. Step-gradient CEC separation of six commonly used diuretics. Conditions: 23.0 cm × 50 μm, CEC Hypersil C18, 3 μm; MeCN/50 mM Na_2PO_4, pH 2.5/H_2O (Initial 40:20:40 for 0–6.50 min, 60:20:20 for 6.50–17.25 min, 40:20:40 for 17.25–25.00 min); 30 kV, 8 bar pressure both ends; 15°C. Detection: UV, 210 nm. Injection: 5 kV/15 s. Sample: 0.2 mg/mL, (1) chlorothiazide, (2) hydrochlorothiazide, (3) chlorthalidone, (4) hydroflumethiazide, (5) bendroflumethiazide, (6) bumetadine. (From Ref. [48], with permission.)

Taylor and Teale have combined gradient CEC with electrospray ionization MS for detection and demonstrated its applicability using thiazide diuretics, corticosteroids, and benzodiazepines [49].

5.2.5 Antibiotics

OT-CEC using etched capillaries bonded with a C18 moiety were prepared and used by Pesek and Matyska for the separation of tetracyclines (50). The results obtained were better than those achieved using polymeric HPLC columns and comparable to those obtained on diol

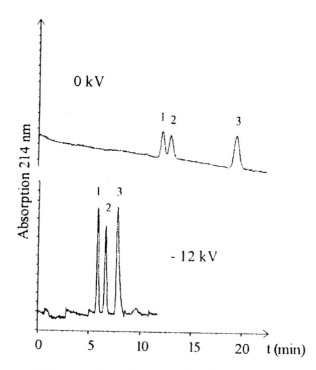

Figure 21. PEC separation of beta-agonists showing the effects of applied voltage. Conditions: 18.0 cm × 100 µm, Nucleosil C8, 5 µm; MeOH/20 mM ammonium acetate, pH 2.9 (50:50). Detection: UV, 214 nm. Sample: (1) terbutaline, (2) fenoterol, (3) clenbuterol. (From Ref. [51], with permission.)

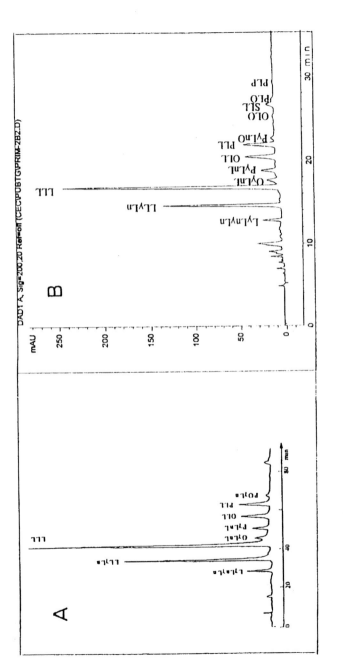

Figure 22. Triglyceride analysis of primrose oil by micro-LC (A) and CEC (B). (A) micro-LC Conditions: 50.0 cm × 320 μm, BioSil C18 HL, 5 μm; MeCN/isopropanol/n-hexane (57:38:5); 5 μL/min; 20°C. Detection: UV, 210 nm. (B) CEC Conditions: 25.0 cm × 100 μm, Hypersil ODS, 3 μm; MeCN/isopropanol/n-hexane (57:38:5) plus 50 mM ammonium acetate; 30 kV; 20°C. Detection: UV, 200 nm. Injection: 10 bar/3 s. (From Ref. [52], with permission.)

columns. The detection limits of 5 to 10 µg/mL were also comparable to both HPLC and CE.

5.2.6 Other Related Applications

Separations of prostaglandin and related impurities, cephalosporin, antibiotic cefuroxime axetil and its E-Z isomers and diastereomers, nonsteroidal anti-inflammatory drugs (NSAIDs), and barbiturates have all been separated using CEC in the reversed phase or mixed reversed phase/cation exchange modes [34, 39]. PEC separations of an antiepileptic drug and its metabolites, 2-phenylethylamines and beta-agonists, have also been reported, as shown in Figure 21 [51]. The use of added pressurized flow overcomes the limitations of low EOF at low pHs.

Sandra et al. have reported enhanced resolution and shorter times for the analysis of triglycerides using CEC vs. micro-LC, as illustrated in Figure 22 [5253]. The CEC methods were applied for the analysis of samples of vegetable and fish oils, margarines, and pharmaceutical formulations using photodiode array (PDA) detection. The RSDs for the retention times were less than 0.5% for electrokinetic injections, while less than 2% was reported for the hydrodynamic injections.

5.3 Drugs of Abuse

The applicability of CEC to analysis of complex mixtures, in which the advantage of improved peak capacity is truly realized, was demonstrated by researchers at the U.S. Drug Enforcement Administration laboratories using cannabinoids [54]. Hashish and marijuana extracts were run along with standards on a C18 column, as illustrated in Figure 23. In order to overcome the concentration sensitivity limitations encountered with capillary techniques, a high-sensitivity UV detection cell was employed, resulting in an eight-fold improvement in signal-to-noise ratio and with minimal loss in resolution, approaching the sensitivity of HPLC.

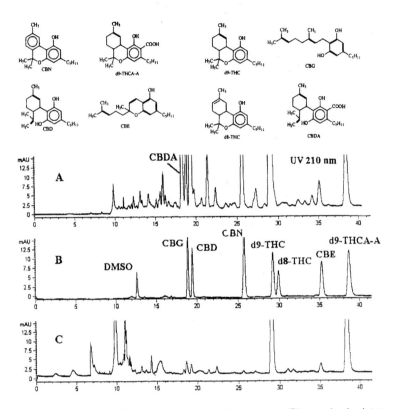

Figure 23. CEC of (A) concentrated hashish extract, (B) standard mixture of cannabinoids, and (C) concentrated marijuana extract. Conditions: 40 (49) cm × 100 µm, Hypersil C18, 3 µm; MeCN/25 mM phosphate, pH 2.57 (75:25); 30 kV, 10 bar pressure both ends. Detection: PDA-UV, 210 nm. Injection: 5 kV/8 s. (From Ref. [54], with permission.)

5.4 Ion Analysis

Tsuda and coworkers reported the separation of various small ions such as sulfite, sulfate, and thiosulfate, along with a simultaneous separation of anions and cations using ion exchange resins and indirect UV detection [55]. The separation of cations is based on their difference in electrophoretic mobilities through the packed bed, whereas anions are separated due to both an ion exchange mechanism and mobility differences on a column packed with anion exchange resin, as shown in Figure 24. Note that these separations were achieved with the use of applied pressure (PEC) to minimize bubble formation and to push the anions past the detector. Mixtures of lanthanide ions were also separated in a similar fashion. The alternate selectivity offered by CEC versus CZE was exploited for the analysis of iodide and iodate, which were constituents of nuclear waste, along with ReO_4^- ions [56]. Reversed polarity was used to sweep the anions past the detector. Although a 20-fold improvement in the limit of detection (2.5×10^{-16} mol for iodide) compared with CZE (4.5×10^{-15} mol) was obtained, the poor reproducibility of the analysis limited successful application of the method.

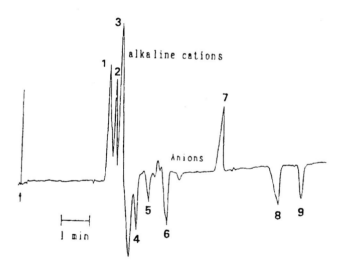

Figure 24. Simultaneous PEC separation of cations and anions with indirect detection. Conditions: 22 (30.4) cm × 50 μm, TSK IC-Anion-SW, 5 μm; MeOH/5 mM phthalic + 5 mM hexamethylenediamine + 15% HEPES, pH 6.8 (10:90); 4 kV. Detection: UV, 236 nm. Sample: (1) Li^+, (2) Na^+, (3) K^+, (4) Cl^-, (5) NO_2^-, (6) NO_3^-, (7) I^-, (8) SO_4^{2-}, (9) ClO_4^-. (From Ref. [55], with permission.)

5.5 Amino Acids

Fujimoto et al. reported the separation of dansyl amino acids on columns filled with linear and cross-linked polyacrylamides [8182]. EOF was achieved by the inclusion of moieties, such as 2-methyl-1-propanesulfonic acid (AMPS) in the polymer network, resulting in a charged column. Here, the separation was based primarily on the restricted migration of the analyte through the gel matrix, owing to the differences in molecular size, as in capillary gel electrophoresis (CGE). This can be considered a CEC separation since there may also be interactions between the analyte and the functional groups on the polymer. Phenylthiohydantoin (PTH) amino acids have been separated

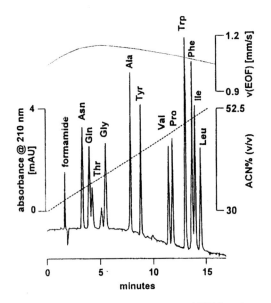

Figure 25. Gradient-elution CEC separation of PTH-amino acids. Conditions: 12.7 (20.7) cm × 50 µm, Zorbax ODS, 3.5-µm packing; mobile phase A: ACN-5 mM phosphate buffer, pH 7.55 (30/70, v/v); B: ACN-5 mM phosphate, pH 7.55 (60/40), 0 to 100% B in 20 min at 0.1 mL/min; 10 kV; 25°C; detection: UV, 210 nm; injection: 1 kV/0.5 s; sample: 30–60 µg/mL, in order of elution: formamide, PTH-asparagine, PTH-tryptophan, PTH-phenyl-alanine, PTH-isoleucine, PTH-leucine. (From Ref. [43], with permission.)

on monolithic ODS columns using acetonitrile/phosphate buffer and packed-bed C18 columns with isocratic and gradient elution [43, 83–84]. Figure 25 illustrates the separation obtained using a linear gradient [43].

5.6 Enantiometric Separations

As with CE, CEC may find a niche in the area of enantiomeric separations. The low volumes of reagents required for the capillary techniques may be of advantage, as chiral phases and additives are expensive. Various approaches have been taken for the CEC separation of enantiomers, combining the experiences of both HPLC and CE. Chiral selectors typically used in HPLC and CE, such as proteins, cellulose-based and Pirkle-type stationary phases, molecularly imprinted polymers (MIPs), and cyclodextrins, have been utilized. However, it should be noted that although higher efficiencies and better resolutions may

Table 3. Enantiomeric CEC Separations Using Molecularly Imprinted Polymers

Compound	Phase	Mode	Conditions	Reference
Phenylalanine, phenyl-alanine anilide, phenylglycine	MIP	CEC	MeCN/HAc/water (90/5/5 or 70/20/10)	57–60
Dansyl leucine, tryptophan	MIP	CEC	MeCN/HAc/water (80/10/10 or 90/5/5)	58–61
Dansyl phenylalanine	MIP	OT-CEC	MeCN/10 mM phosphate, pH 7.0 (80/20)	62
Propanolol, meto-prolol, prenalterol, atenolol, pindolol	MIP	CEC	MeCN/4 M acetate, pH 3.0 (80/20)	63, 65
Ropivacaine, mepiva-caine, bupivacaine	MIP	CEC	MeCN/25 mM phos-phate, pH 3.0 (80/20)	64, 65

Table 4. Enantiomeric Separations Using Protein Phases

Compound	Phase	Mode	Conditions	Reference
Benzoin, barbiturates, ifosfa- mide. cyclophosphamide, disopyramide, β-agonists	AGP	CEC	Propanol/phosphate, pH and composition varied	68
Benzoin, temazepam	HSA	CEC	Propanolol/4 mM phosphate, pH 7.0, composition varied	69

AGP = α-1-acid glycoprotein; HSA = human serum albumin.

be obtained versus HPLC, the problem of detection sensitivity still persists, as in CE. The majority of the literature demonstrates selectivity, but the research has not progressed to a stage with limits of detection in the range of practical applications for the pharmaceutical industry. Tables 3, 4, and 5 summarize some of the approaches to CEC separation of chiral analytes.

5.6.1   Molecularly Imprinted Polymers

The use of MIPs for the CEC separation of chiral analytes is analogous to the use of these phases in HPLC. The MIPs are prepared by copolymerization of a cross-linker and functional monomers in the presence of a template. Subsequent washing of the polymer reveals recognition sites containing functionalities positioned to complement those of the template molecules. For CEC, the EOF is generated by the functional groups of the polymer, such as the carboxyl moiety. Both columns packed conventionally with MIP particles and in situ–generated MIP monoliths have been used. The primary advantage of this approach is that a customized phase with predetermined selectivity for the analyte of interest can be prepared. Typical separations are carried out using an electrolyte with high levels of the organic modifier and a buffer at low pH.

Table 5. Enantiomeric Separations Using Cyclodextrin Chiral Selectors

Compound	Phase	Mode	Conditions	Reference
Salsoninol	ODS	CEC	20 mM phosphate, pH 3.0 + 12 mM βCD + 5 mM sodium heptanesulfonate	70
Hexobarbital, benzoin, dansyl amino acids, dinitrophenyl amino acids	βCD	CEC	MeOH/phosphate (15/85); MeOH/15 mM TEAA, pH 4.7 (15/85)	71
Chlorthalidone, mianserin	ODS	CEC	MeCN/1 mM phosphate, pH 6.5 (15/85), with 10 mM HPβCD	72
	HPβCD	CEC	MeCN/5 mM phosphate, pH 6.5, composition varied	
Barbiturates, aromatic alcohols	βCD	OT-CEC	MeCN or MeOH/borate-phosphate, pH 7.0	73, 74
Barbiturates, benzoin, α-methyl-α-phenyl-R-succinnimide, glute-thimide, MTH-proline, methyl mandelate	PMβCD	PEC	MeOH/5 mM phosphate, pH 7.9 (80/20)	75

βCD = β-cyclodextrin; HPβCD = hydroxypropyl-β-cyclodextrin, OT-CEC = open tubular capillary electrochromatography; PMβCD = permethyl-β255-cyclodextrin.

Lin and coworkers produced polymers imprinted with various amino acids and studied the effect of eluent composition, polymer particle size, temperature, and applied field [5761]. Polymers imprinted with phenylalanine and phenylalanine anilide were packed into a capillary and tested for separation of the template molecules and other related aromatic amino acids. Although the efficiencies were improved compared to HPLC, the retention times were longer. Columns prepared by polymerization in the capillary have also been evaluated [61]. Remcho and Tan used OT-CEC for the separation of dansyl phenylalanine [62]. A thin layer of the MIP was coated onto the walls of the fused silica capillary. An efficiency of 248,600 plates/m was reported for the nonimprinted enantiomer; however, the imprinted analyte yielded only 8000 plates/m. Baseline resolution was achieved, but the retention times were almost 30 minutes. This may be due to incomplete method optimization.

Nilsson et al. have prepared monolithic columns by in situ polymerization using β-adrenergic antagonists and local anesthetics as templates [63–67]. The effect of porogen, the amount of monomer, and the choice of template for the resulting columns were studied. The separations reported are fast, with detection of 1%

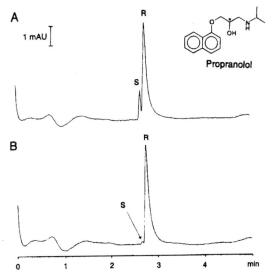

Figure 26. Enantiomeric separation of propranolol on a molecularly imprinted polymer column. Conditions: 35 cm × 75 μm, (R)-propranolol imprinted column; MeCN/4 mM acetate, pH 3 (80:20); 15 kV; 60°C. Detection: UV, 214 nm. Injection: 5 kV/5 s. Samples: (A) 9:1 mixture of (R)- and (S)-propranolol, (B) 99:1 mixture of of (R)- and (S)-propranolol. (From Ref. [63], with permission.)

tiomer for propranolol, as reported in Figure 26. The propranolol-imprinted column was also used to separate other β-antagonists, such as prenalterol, atenolol, and pindolol, with varying degrees of success. A ropivacaine-imprinted polymer column was shown to also discriminate between enantiomers of its structural analogs, mepivacaine and bupivacaine.

5.6.2   Protein Phases

Lloyd et al. have reported the use of columns packed with 5 μm α-1-acid glycoprotein (AGP) particles for the separation of a variety of compounds typically separated on the same phase using HPLC [6869]. Although neutral and cationic enantiomeric compounds could be resolved, anionic analytes were either unresolved or not eluted from the capillary, since they migrated against the EOF. Columns packed with human serum albumin (HSA) immobilized onto silica particles have also been evaluated for the separation of benzoin and temazepam. The separation efficiencies were comparable with HPLC.

Another approach is to pack the columns with the protein immobilized in a gel. Nilsson et al. used capillaries filled with bovine serum albumin (BSA) and cellulase (CBH I) gel to separate β-adrenergic antagonists [65]. The use of gel-packed capillaries for CEC is very similar to capillary gel electrophoresis (CGE), except that in CGE, the gel is not being used to generate EOF, whereas in CEC, it presumably does that as well as perform the desired separations. Nevertheless, there is a striking similarity, at times, in the operational conditions of CGE and CEC, especially when gels are used in both formats.

5.6.3   Cyclodextrins

Cyclodextrins coated onto columns, bonded onto silica beads and as buffer additives, have been used routinely for the separation of enantiomers by HPLC, HPCE, and GC. In a manner analogous to HPLC, Deng et al. used β-cyclodextrin (β-CD) as a buffer additive for CEC and PEC separations of enantiomers of the neurotoxin salsolinol, a hydrophilic amine [70]. A reversed-phase ODS column with phosphate buffer plus β-CD and sodium 1-heptanesulfonate was employed. Capillaries packed with β-CD stationary phase were used by Li and Lloyd for the separation of benzoin, hexobarbital, and a series of dinitrophenyl- and dansyl amino acids [71]. For the anionic amino acids, triethylammonium acetate buffer at pH 4.7 was used to reverse the EOF. The separation efficiencies, however, were less than ideal, and peak asymmetry factors were high.

A comparison of the use of hydroxypropyl-β-cyclodextrin (HPβCD) as a buffer additive and as a chiral stationary phase (bonded onto silica particles) for CEC was provided by Lelievre et al. [72] A 5-µm Cyclobond I 2000 RSP material, conventionally used in HPLC, was employed for the separation of the drugs chlorthalidone and mianserin. Baseline resolution was achieved for chlorthalidone enantiomers using acetonitrile/phosphate buffer as the electrolyte, with the relative standard deviations for the EOF velocity and retention time of 1% and 1.7%, respectively, for the first enantiomer. When the HPβCD was used as a buffer additive, the overall analysis time was longer and the resolution was lower (92 min, Rs = 1.4 for chlorthalidone) versus the chiral stationary phase (39 min, Rs = 1.7; 49 min, Rs = 3.4). However, higher efficiencies (22,000) were obtained with the buffer additive over the packed capillaries (~10,000).

Schurig and coworkers reported on the use of fused silica capillaries coated with Chirasil-Dex, permethyl-β-cyclodextrin (PMβCD) covalently linked to dimethylpolysiloxane, for the resolution of various chiral alcohols and barbituric acids [73-74]. Comparisons were made using the same column in the OT-CEC and OT-micro-LC modes for the separation of analytes using similar electrolytes, as summarized in Table 6. Improved resolutions and efficiencies were obtained for most analytes; the retention times; however, were longer for some of the analytes. This again may simply be due to not-quite optimized conditions. When columns packed with the PMβCD-modified silica stationary phase were examined for barbiturates, higher efficiencies were obtained versus micro-LC using the same columns with comparable

Table 6. Enantiomeric Separations by OT-LC and OT-CEC on Chirasil-Dex

Analyte	Mode	t_{r1}/t_{r2} (min)	N_1/N_2	R_S	T (°C)	Voltage (kV) (OTEC), pressure (bar) (OTLC)	Organic modifier
5-[2-Propyl]-5-n-propyl barbituric acid (a)	OTEC	12.67/15.71	270/1600	2.4	25	30	—
	OTLC	5.82/6/64	1800/1000	1.2	25	0.5	—
5-[1-Cyclohexenyl]-1,5-dimethyl barbituric acid (a)	OTEC	21.90/24.80	65,500/13,600	4.9	20	30	
	OTLC	7.90/9.20	5500/2900	2.4	20	0.15	
5-Ethyl-5-[1-methylbutyl] barbituric acid (s)	OTEC	14.41/16.38	1800/1700	1.3	20	30	
	OTLC	3.81/4.43	2000/1200	1.4	20	0.3	
5-Ethyl-5-phenyl barbituric acid (d)	OTEC	26.20/30.39	50,000/18,800	6.0	30	30	—
	OTLC	27.85/33.07	8400/5000	3.4	30	0.06	
5-Ethyl-5-n-propyl barbituric acid (a)	OTEC	11.35/14.42	8900/3400	4.2	20	30	—
	OTLC	31.50/34.06	4600/2700	1.1	20	0.1	
6-Chloro-α-methylcarbazole-2-acetic acid (b)	OTEC	36.16/39.09	47,900/18,300	4.3	30	30	
	OTLC	6.40/8.07	1700/700	1.8	30	0.8	
1-Phenyethanol (a)	OTEC	12.80/13.74	47,400/23,900	3.2	30	30	=
	OTLC	24.80/26.50	24,700/22,200	2.5	30	0.07	=

Table 6. (Continued)

Analyte	Mode	t_{r1}/t_{r2} (min)	N_1/N_2	R_S	T (°C)	Voltage (kV) (OTEC), pressure (bar) (OTLC)	Organic modifier
1-Phenyl-1-propanol (a)	OTEC	26.60/30.79	5600/2700	2.2	30	25	II
	OTLC	41.20/47.44	3000/2600	1.9	40	0.055	II
α-Methyl-2,3,4,5,6-pentafluorbenzyl-alcohol (a)	OTEC	23.10/25.28	10,900/3500	1.7	20	30	II
	OTLC	21.40/24.52	12,400/3900	2.7	20	0.45	III
1-(2-Naphthyl)-ethanol (a)	OTEC	22.40/25.00	19,400/12,700	3.4	60	30	II
	OTLC	16.00/18.95	4400/4500	2.8	60	0.12	II
1-(p-Biphenylethanol) (a)	OTEC	33.30/34.20	42,400/16,900	1.1	60	30	IV
	OTLC	35.30/37.25	10,600/4400	1.1	60	0.075	IV
5-Ethyl-5-n-propyl barbituric acid (c)	OTEC	6.83/7.86	7100/3000	2.3	20	30	
	OTLC	3.68/4.73	1100/500	1.6	20	0.6	
5-Ethyl-5-phenyl barbituric acid (c)	OTEC	5.86/7.54	12,200/3900	4.8	20	30	
	OTLC	4.73/6.31	1000/500	1.9	20	0.6	
1-Phenylethanol (a)	OTEC	12.80/13.74	47,400/23,900	3.2	30	30	II
	OTLC	24.80/26.50	24,700/22,200	2.5	30	0.07	II

Conditions: Column: (a) 50 μm i.d. × 0.8 m effective length (0.95 m total length); (b) 50 μm i.d. × 0.6 m effective length (0.7 m total length); (c) 25 μm i.d. × 0.5 m effective length (0.62 m total length); (d) 50 μm i.d. × 0.88 m effective length (0.98 m total length). Buffer for all measures: borate–phosphate buffer 20 mM (pH 7). Modifier: (I) buffer/acetonitrile 90/10 (v/v); (II) buffer/methanol 97/3 (v/v); (III) buffer/methanol 91/9 (v/v); (IV) buffer/methanol 62/38 (v/v).

From Ref. [74], with permission.

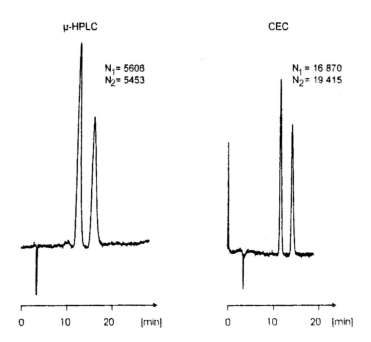

Figure 27. Enantiomeric separation of mephobarbital by (a) micro-LC and (b) PEC. Conditions: 23.5 (40) cm × 100 µm, PB-β-CD silica, 5 µm, MeOH/5 mM phosphate, pH 7.0 (20:80); (a) 20 kV with 10 bar applied pressure; (b) 140 bar. Detection: UV, 230 nm. Injection: 100 bar/5–10 s. (From Ref. [75], with permission.)

elution times, as illustrated in Figure 27 [75]. The electrochromatography in this case was performed with the assistance of pressure (i.e., as PEC).

5.6.4 Brush-Type Phases

By far, the most encouraging results for enantiomeric separations by CEC have been obtained with the brush-type, or Pirkle, chiral stationary phases that have rapid mass transfer characteristics. Capillaries packed with (*S*)-naproxen–derived and (*3R,4S*)-Whelk-O1–based phases immobilized onto 3-µm silica support were used with MES buffer/acetonitrile as the electrolyte [76]. A variety of neutral compounds were sepa-

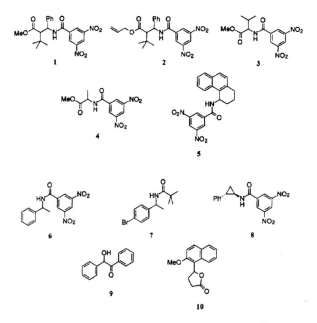

Table 7. CEC Enantioseparations of Compounds 1–5 on (S)-Naproxen–Derived Chiral Stationary Phase and 6–10 on (3R,4S)-Whelk-O 1 CSP

Compound	k_1	α	N_1/m	N_2/m	R_S
1	0.77	1.90	195,000	68,000	13.47
2	0.86	1.78	173,000	49,000	11.30
3	0.53	1.49	196,000	114,000	8.02
4	0.51	1.33	170,000	121,000	5.61
5	1.28	2.80	121,000	24,000	16.86
6	0.61	1.59	180,000	154,000	11.48
7	0.34	3.82	200,000	160,000	30.95
8	0.75	1.29	157,000	145,000	6.29
9	0.26	1.23	176,000	189,000	2.63
10	0.54	1.54	182,000	170,000	10.14

Conditions: Compounds 1–5, 25 mM MES, pH 6.0/acetonitrile (1:3:5), 25 kV, 38.0 cm total capilary length (29.5 cm effective length); compounds 6–10, same except for 39.3 total capillary length (30.5 cm effective length).
From Ref. [76], with permission.

rated, most in under 10 minutes, with resolutions ranging from 2.6 to 31. Very high efficiencies were obtained for both of the enantiomers, as described in Table 7.

Monolithic columns incorporating a valine-based, brush-type selector have also been prepared, yielding efficiencies of 61,000 and 49,500 plates for the enantiomers of N-(3,5-dinitrobenzoyl)leucine under optimized conditions. Analysis times under 6 minutes and a resolution of 2.0 were obtained, as shown in Figure 2.13 (p. 54) [77].

5.6.5 Antibiotics

Columns packed with the glycopeptide antibiotic vancomycin coated onto 5-μm silica particles have also been explored for CEC applications

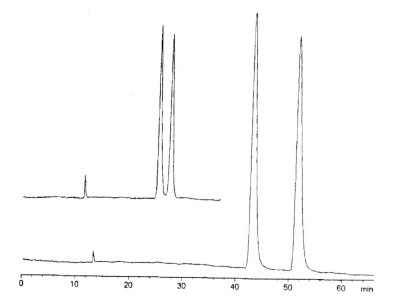

Figure 28. CEC enantioseparation of warfarin. Conditions: 40 cm × 100 μm, vancomycin CSP, 5 μm; MeCN-0. 1% TEAA, pH 5 (20:80); 30 kV; 20°C. Detection: UV, 200 nm. Injection: 10 kV/1 s. Inset: MeCN/0. 1% TEAA, pH 5 (30:70). From Ref. [78], with permission.

[78]. Warfarin and hexobarbital enantiomers were separated using triethylamine acetate buffer and acetonitrile as the mobile phase. Figure 28 shows the separation of warfarin with resolution of 2.7, selectivity of 1.28, and plate counts of 13,300 for the later-eluting component. Increasing the organic concentration resulted in decreased analysis times; however, enantioselectivity was compromised. For hexobarbital, efficiencies of 38,750 plates/m were obtained with CEC compared with 27,000 plates/m using a chirobiotic vancomycin HPLC column. Teicoplanin, a macrocyclic antibiotic, has also been used as a chiral selector in CEC for the separation of ibuprofen and phenylalanine enantiomers [79].

5.6.6   Other Phases

The vast selection of chiral selectors employed in HPLC and CE separations can all be utilized with CEC. Francotte and Jung have made wall-coated capillaries with cellulose-derived phases, 3,5-dimethyl-phenylcarbamoyl cellulose (DMPCC) and *para*-methylbenzoyl cellulose (PMBC) for OT-CEC of basic and neutral pharmaceuticals [80]. Although the columns could be manufactured reproducibly, their lifetimes were short, usually less than a hundred injections.

5.7   Conclusions

It is apparent that there remains a large amount of interest in the application of CEC to small-molecule pharmaceuticals, just as there was for CE and micellar electrokinetic chromatography (MEKC) [8384]. There are literally thousands of HPLC applications that could be transferred to CEC, provided that there are practical justifications for such transfers and expenditure of funds. Although resolutions may be somewhat different than by HPLCat times even betterthere are lingering problems for the routine, practical application of CEC to this class of molecules. Commercial CEC capillaries are a bit more expensive than current HPLC columns, they are generally less tolerant of dirty samples such as biofluids, and they may require more sample preparation and cleanup of crude samples.

CEC does not quite provide the sample loadability and therefore limits of detection, in comparison with most instances of advanced HPLC approaches and techniques. Although resolutions as well as plate counts may be somewhat better for many CEC applications of small molecules, that may not matter when there are only a few analyte peaks of interest, as in most pharmaceutical analyses. What is the real utility of generating more plates per meter if there are only 3"Symbol"-5 analyte peaks that need resolving? HPLC can still do that more easily and reproducibly, and probably at much lower cost. This may be the real sticking point in all future applications of CEC outside of academia or research institutes.

Why then would a pharmaceutical firm utilize CEC for routine sample analyses, when HPLC has served and continues to serve so well? The ability of CEC to provide shorter overall analysis timesseven in the isocratic mode, because of its increased peak capacity and narrower peak shapesmay be important for in-process sample analyses. This is where turnaround time is critical, since such testing can hold up further processing of bulk drugs and their formulations.

As in CE, investigations into chiral CEC applications deserve further efforts, since current HPLC techniques are expensive, time consuming, and do not always yield ideal resolutions of the chiral analytes. Compared with HPLC, chiral CEC offers improved resolutions, higher efficiencies, improved selectivity, suitable quantitation, accurate enantiomeric excess determinations, and everything else desired in a good chiral separation. However, there are, at present, no commercial chiral CEC capillaries on the market. These will need to be more widely available, more rugged, and at lower cost to win over chiral HPLC users.

References

1. M. M. Robson, M. G. Cikalo, P. Myers, M. R. Euerby, and K. D. Bartle. *Journal of Microcolumn Separations, 9,* 357 (1997).

2. M. G. Cikalo, K. D. Bartle, M. M. Robson, P. Myers, and M. R. Euerby. *The Analyst, 123,* 87R (1998).

3. M. R. Euerby, D. Gilligan, C. M. Johnson, S. C. P. Roulin, P. Myers, and K. D. Bartle. *Journal of Microcolumn Separations, 9,* 373 (1997).

4. M. M. Dittmann and G. P. Rozing. *Journal of Chromatography A, 744,* 63 (1996).

5. C. Yan, R. Dadoo, H. Zhao, and R. N. Zare. *Analytical Chemistry, 67,* 2026 (1995).

6. J. Ding and P. Vouros. *Analytical Chemistry, 69,* 279 (1997).

7. J. Ding, J. Szeliga, A. Dipple, and P. Vouros. *Journal of Chromatography A, 781,* 327 (1997).

8. J. Ding and P. Vouros. *American Laboratory, 15* (June, 1998).

9. H. Engelhardt, L. Stefan, and F. T. Hafner. *American Laboratory, 40* (April, 1998).

10. R. Dadoo, R. N. Zare, C. Yan, and D. S. Anex. *Analytical Chemistry, 70,* 4787 (1998).

11. J. H. Knox and I. H. Grant. *Chromatographia, 32,* 317 (1991).

12. S. Ludtke, T. Adam, and K. K. Unger. *Journal of Chromatography A, 786,* 229 (1997).

13. C. G. Bailey and C. Yan. *Analytical Chemistry, 70,* 3275 (1998).

14. C. Ericson, J. L. Liao, K. Nakazato, and S. Hjerten. *Journal of Chromatography A, 767,* 33 (1997).

15. J. L. Liao, N. Chen, C. Ericson, and S. Hjerten. *Analytical Chemistry, 68,* 3468 (1996).

16. E. C. Peters, M. Petro, F. Svec, and J. M. Frèchét. *Analytical Chemistry, 69,* 3646 (1997).

17. E. C. Peters, M. Petro, F. Svec, and J. M. Frèchét.. *Analytical Chemistry, 70,* 2288 (1998).

18. E. C. Peters, M. Petro, F. Svec, and J. M. Frèchét. *Analytical Chemistry, 70,* 2296 (1998).

19. J. Z. Tan and V. T. Remcho. *Journal of Micro. Separations, 10,* 99 (1998).

20. Y. Guo and L. Colón. *Journal of Micro. Separations, 7,* 485 (1995).

21. Y. Gou and L. Colón. *Analytical Chemistry, 67,* 2511 (1995).

22. P. Narang and L. Colón. *Journal of Chromatography A, 773,* 65 (1997).

23. A. Maruska and U. Pyell. *Chromatographia, 45,* 229 (1997).

24. A. Maruska and U. Pyell. *Journal of Chromatography A, 782*, 167 (1997).
25. K. W. Whitaker and M. J. Sepaniak. *Electrophoresis, 15*, 1341 (1994).
26. V. Lopez-Avila, J. Benedicto, and C. Yan. *Journal of High Resolution Chromatography, 20*, 615 (1997).
27. Y. Zhang, W. Shi, L. Zhang, and H. Zou. *Journal of Chromatography A, 802*, 59 (1998).
28. F. Moffatt, P. Chamberlain, P. A. Cooper, and K. M. Jessop. *Chromatographia, 48*, 481 (1998).
29. N. C. Gillott, M. R. Euerby, C. M. Johnson, D. A. Barrett, and P. N. Shaw. *Anal. Commun., 35*, 217 (1998).
30. I. S. Lurie, T. S. Conver, and V. L. Ford. *Analytical Chemistry, 70*, 4563 (1998).
31. W. Wei, G. A. Luo, G. Y. Hua, and C. Yan. *Journal of Chromatography A, 817*, 65 (1998).
32. N. W. Smith and M. B. Evans. *Chromatographia, 41*, 197 (1995).
33. W. Wei, G. Luo, and C. Yan. *American Laboratory, 20C* (January, 1998).
34. M. R. Euerby, C. M. Johnson, and K. D. Bartle. *LC-GC International, 39* (January, 1998).
35. T. Eimer, K. K. Unger, and T. Tsuda. *Fresenius Journal of Analytical Chemistry, 352*, 649 (1995).
36. K. D. Altria, N. W. Smith, and C. H. Turnball. *Journal of Chromatography B, 717*, 341 (1998).
37. J. Wang, D. E. Schaufelberger, and N. A. Guzman. *Journal of Chromatographic Science, 36*, 155 (1998).
38. M. R. Euerby , C. M. Johnson, K. D. Bartle, P. Myers, and S. C. P. Roulin. *Anal. Commun., 33*, 403 (1996).
39. N. W. Smith and M. B. Evans. *Chromatographia, 38*, 649 (1994).
40. S. J. Lane, R. Boughtflower, C. Paterson, and T. Underwood. *Rapid Communications in Mass Spectrometry, 9*, 1283 (1995).
41. S. J. Lane, R. Boughtflower, C. Paterson, and M. Morris. *Rapid Communications in Mass Spectrometry, 10*, 733 (1996).
42. D. B. Gordon, G. A. Lord, and D. S. Jones. *Rapid Communications in Mass Spectrometry, 8*, 544 (1994).
43. C. G. Huber, G. Choudhary, and C. Horvath. *Analytical Chemistry, 69*, 4429 (1997).

44. R. M. Seifar, W. T. Kok, J. C. Kraak, and H. Poppe. *Chromatographia, 46,* 131 (1997).

45. R. M. Seifar, J. C. Kraak, W. T. Kok. and H. Poppe. *Journal of Chromatography, 808,* 71 (1998).

46. M. R. Taylor , P. Teale, S. A. Westwood, and D. Perrett. *Analytical Chemistry, 69,* 2554 (1997).

47. D. A. Stead, R. G. Reid, and R. B. Taylor. *Journal of Chromatography A, 798,* 259 (1998).

48. M. R. Euerby, D. Gilligan, C. M. Johnson, and K. D. Bartle. *The Analyst, 122,* 1087 (1997).

49. M. R. Taylor and P. Teale. *Journal of Chromatography, A, 768 ,* 89 (1997).

50. J. J. Pesek and M. T. Matyska. *Journal of Chromatography A, 736,* 313 (1996).

51. T. Eimer, K. K. Unger, and J. van der Greef. *Trends in Analytical Chemistry, 15,* 463 (1996).

52. P. Sandra, A. Dermaux, V. Ferraz, M. M. Dittmann, and G. Rozing. *Journal of Microcolumn Separations, 9,* 409 (1997).

53. A. Dermaux, P. Sandra, M. Ksir, and K. F. F. Zarrouck. *Journal of High Resolution Chromatography, 21,* 545 (1998).

54. I. S. Lurie, R. P. Meyers, and T. S. Conver. *Analytical Chemistry, 70,* 3255 (1998).

55. D. Li, H. H. Knobel, S. Kitagawa, A. Tsuji, H. Watanabe, M. Nakshima, and T. Tsuda. *Journal of Microcolumn Separations, 9,* 347 (1997).

56. V. T. Remcho. *Journal of Chromatography B, 695,* 169 (1997).

57. J. M. Lin, T. Nakagama, K. Uchiyama, and T. Hobo. *Journal of Liquid Chromatography & Related Technologies, 20,* 1489 (1997).

58. J. M. Lin, T. Nakagama, K. Uchiyama, and T. Hobo. *Biomedical Chromatography, 11,* 298 (1997).

59. J. M. Lin, T. Nakagama, X. Z. Wu, K. Uchiyama, and T. Hobo. *Fresenius Journal of Analytical Chemistry, 357,* 130 (1997).

60. J. M. Lin, K. Uchiyama, and T. Hobo. *Chromatographia, 47,* 625 (1998).

61. J. M. Lin, T. Nakagama, K. Uchiyama, and T. Hobo. *Journal of Pharmac. Biomedical Analysis, 15,* 1351 (1997).

62. Z. J. Tan, and V. T. Remcho. *Electrophoresis, 19,* 2055 (1998).

63. L. Schweitz, L. I. Andersson, and S. Nilsson. *Analytical Chemistry, 69,* 1179 (1997).
64. L. Schweitz, L. I. Andersson, and S. Nilsson. *Journal of Chromatography A, 792,* 401 (1997).
65. S. Nilsson, L. Schweitz, and M. Petersson. *Electrophoresis, 18,* 884 (1997).
66. L. Schweitz, L. I. Andersson, and S. Nilsson. *Journal of Chromatography A, 817,* 5 (1998).
67. O. Bruggemann, R. Freitag, M. J. Whitcombe, and E. N. Vulfson. *Journal of Chromatography A, 781,* 43 (1997).
68. S. Li and D. K. Lloyd. *Analytical Chemistry, 65,* 3684 (1993).
69. D. K. Lloyd, S. Li, and P. Ryan. *Journal of Chromatography A, 694,* 285 (1995).
70. Y. Deng, J. Zhang, T. Tsuda, P. H. Yu, A. A. Boulton, and R. M. Cassidy. *Analytical Chemistry, 70,* 4586 (1998).
71. S. Li and D. K. Lloyd. *Journal of Chromatography A,* 666, 321 (1994).
72. F. Lelievre, C. Yan, R. N. Zare, and P. Gareil. *Journal of Chromatography A, 723,* 145 (1996).
73. S. Mayer and V. Schurig. *Journal of High Resolution Chromatography, 15,* 129 (1992).
74. H. Jakubetz, H. Czesla, and V. Schurig. *Journal of Microcolumn Separations, 9,* 421 (1997).
75. D. Wistuba, H. Czesla, M. Roeder, and V. Schurig. *Journal of Chromatography A, 815,* 183 (1998).
76. C. Wolf, P. L. Spence, W. H. Pirkle, E. M. Derrico, D. M. Cavender, and G. P. Rozing. *Journal of Chromatography A, 782,* 175 (1997).
77. E. C. Peters, K. Lewandowski, M. Petro, F. Svec, and J. M. J. Frechet. *Anal. Commun., 35,* 83 (1998).
78. A. Dermaux, F. Lynen, and P. Sandra. *Journal of High Resolution Chromatography, 21,* 575 (1998).
79. J. H. Miyawa and M. S. Alesandro. *LC-GC, 16,* 36 (1998).
80. E. Francotte and M. Jung. *Chromatographia, 42,* 521 (1996).
81. C. Fujimoto, J. Kino, and H. Sawada. *Journal of Chromatography A, 716,* 107 (1995).
82. C. Fujimoto. *Analytical Chemistry, 67,* 2050 (1995).
83. R. Asiaie, D. Huang, D. Farnan, and C. Horvath. *Journal of Chromatography A, 806,* 251 (1998).

84. C. G. Huber, G. Choudhary, and C. Horvath. *Analytical Chemistry, 69*, 4429 (1997).

6 Biopolymers

B iopolymers, especially proteins and peptides, can be separated based on their differing degrees of hydrophobicity in reversed-phase (RP) HPLC modes, their pI differences in ion-exchange chromatography (IEC) or chromatofocusing modes, and differences in their solubilities in aqueous solution in hydrophobic interaction chromatography (HIC) HPLC modes [1–10].

At times, it may prove appealing to utilize a mixed-mode type packing in CEC, such as a combined cation exchange chromatography (CIEC) and RP support to separate charged species, especially proteins. The ion-exchange groups can provide enhanced EOF compared with a purely RP stationary phase. By carefully considering buffer, pH, and organic content, one could even alternate the separation mechanism from RP to IEC or a combination of the two, as long as the analytes eventually are separated and eluted. There are, however, pitfalls in such an approach using CEC: potentially charged species, such as peptides, may undergo adsorption onto the ion-exchange sites under RP conditions and never elute. Buffer/elution conditions must be selected that permit resolution and elution in a reasonable time frame while permitting both enhanced EOF and true partitioning from the IEC sites, as opposed to irreversible attachment.

Other than for capillary gel electrophoresis (CGE), which effectively has a packing (gel) in the capillary and operates under an electrophoretic migration force (EPF) or electrophoretic mobility, HPCE is a truly electrophoretic migration/separation approach. The ability to utilize HPLC stationary-mobile phase interactions, together with electrophoretic mobility for charged biopolymers and to affect differences in EPF and partitioning in RP (pH and organic modifier effected) or pI (pH affected) in IEC modes, should prove to be significant attributes of the newer CEC approaches.

Separation modes aside, the remaining issue is good detectability, which must be addressed. Detectability is affected by preconcentration at the head of the capillary column, stacking, prefocusing, wider (bubble type) capillaries, Z-shaped capillaries, internal reflectance UV cells, and other instrumental or capillary techniques.

6.1 Potential of CEC

By and large, CEC appears to offer many of the desirable traits needed for successful biopolymer separations (Table 1). That is, one can utilize both stationary-mobile phase interactions, just as with RP and IEC HPLC modes, but with the addition of an electrophoretic component or resolving force not available in HPLC. HPCE can also provide an EPF that HPLC does not provide; however, it does not have the inherent stationary-mobile phase partitioning or ion-exchange interactions that are so well demonstrated for HPLC.

The real advantage that CEC can provide has to do with its (1) more symmetrical and very narrow (high-efficiency) peak shapes, resulting in higher peak capacities and (2) improved resolution with the use of counteracting flow EOF and EPF. The small differences in mobilities between similar peptides, proteins, nucleic acids, and other biopolymer classes, even perhaps carbohydrates, should prove a serious advantage over just HPLC or HPCE alone, when these can be utilized

Table 1. Applicability of CEC to Biopolymer Separations

Goal	Effectiveness of CEC
Narrow, sharp eluting peaks with good peak symmetry (unit asymmetry value = 1)	Demonstrated: analyte dependent
High peak capacity per capillary length	Demonstrated
High speed of separation and analysis	Demonstrated
Ease of quantitation	Not yet demonstrated
Ease of method development/optimization	Not yet demonstrated
Ease of interfacing with MS	Demonstrated
Trace level detection	Not yet demonstrated
Ability to separate variants of proteins and other biopolymers	Case dependent

at random in the very same capillary in CEC. In general, the judicious use of applied voltage, pH, organic content, and stationary phase in CEC should provide improved separations over the currently demonstrated capability of HPLC or HPCE alone.

CEC, however, is unlikely to replace other, established techniques, just as supercritical fluid chromatography (SFC) and HPCE never really replaced major applications of GC or HPLC. Unless and until certain other desirable features of GC and HPLC are also offered by CEC, users of GC/HPLC will not abandon long-established, well-optimized, and functioning methods, just because CEC may provide improved peak shapes, plate counts, efficiencies, peak capacity, and resolutions.

The acceptance of a newer method of analysis also depends on ease of use, repeatability and reproducibility, method transfer, method validation, method robustness and ruggedness, ease of operation and training, and, in the case of CEC, availability of commercially stable and robust capillaries with a wide variety of packings.

In addition, it must be acknowledged that very well established approaches such as HPLC are, from a practical viewpoint, difficult to dislodge. Although CEC is able to solve very complex separations of mixtures that contain too many compounds for conventional HPLC or even HPCE modes—and there may be other applications or samples for which CEC proves superior—it is obvious that current users of HPLC continue to derive suitable and practical separations of biopolymers, under both analytical and preparative conditions. For the foreseeable future, CEC may remain largely a niche technique.

In reviewing the CEC biopolymer literature—very little of which predates 1994—we examined how the authors chose their experimental conditions, how they then went about optimizing those conditions, whether or not the final conditions really demonstrated optimal use, and whether or not any real samples were ever assayed. We found that researchers generally chose packing materials from commercially available capillaries or packed their own, presumably with whatever packings were available. While this approach to capillary packing selection may be practical, it is often less than ideal and not very scientific. Future research in CEC will be well served by rational method development and a more rigorous approach to choosing and reporting initial CEC conditions.

We also note that the CEC literature on biopolymers has tended to focus on proteins and peptides. Far fewer papers have appeared that deal with carbohydrates, lipids, or nucleic acids.

6.2 Stationary Phase and Packing Considerations

Much of the literature on CEC for biopolymers is devoted to CEC–electrospray ionization (ESI)–MS, with only a little focusing on direct CEC–UV/fluorescence detection (FL) methods. Much of the research has evolved from the use of commercially available, prepacked capillaries, such as C-18 or ion-exchange, or a mixed mode containing both ion-exchange and RP. Very little research has actually attempted to develop new phases specifically for biopolymers.

When CEC for biopolymers is used in conjunction with MS, the CEC conditions never really need to be fully optimized, since the MS accomplishes additional resolution and specific identification, as needed. The specific mobile-phase conditions in CEC-MS may be quite different from those for CEC-UV/FL, just as in CE or HPLC, and thus optimization of CEC-MS conditions will be somewhat different from that for CEC-UV/FL. As for LC-MS, this would include the use of volatile organic solvents and organic buffers, low flow rates, no void volume or loss of resolution in the CEC-MS interface, and the usual interfacing requirements already developed and optimized for CE-MS [11-25].

Few absolute quantitations for biopolymers have been reported. Most of the literature is qualitative by nature, simply demonstrating suitable, if not fully optimized, experimental conditions that provide evidence for the presence of certain biopolymers and their resolution from other components in a particular sample, or percent areas. More true quantitation results (amounts present) for biopolymers are needed.

Successful CEC packings for biopolymers must meet certain requirements. Depending on the ionic characteristics of the biopolymers, ideal packings are pH dependent and are either RP or IEC, or a combination of both [26-36]. Size exclusion packings in CEC have been applied mainly to synthetic organic polymers and much less to biopolymers [37-40]. Regardless of which packing is actually utilized, it must contain a stationary (bonded, not coated) phase that can successfully interact with the biopolymers, as in RP-HPLC, and ideally prevent any unwanted silanol interactions with the underlying silica or ionic sites. It should also provide additional or programmable EOF besides that from the uncoated fused-silica capillary walls. An ideal packing combines a CIEC material with RP to allow separations that are based on RP (hydrophobicity) alone, a combination of RP and IEC, or just IEC alone, and that are mobile phase (buffer) dependent. Whatever the mode, the packing should prevent unwanted interaction of the biopolymer (e.g., peptide amino groups) with the support, such as amine-silanol hydrogen bonding in RP-HPLC for amine-containing analytes (e.g., pharmaceuticals, peptides, and proteins).

In open tubular CEC (OT-CEC) applications, a coating is applied on the inner surface of the capillary as in capillary GC. In packed-bed CEC, the packing is in the capillary. Methods may employ CEC or PEC. There is also electro-HPLC, which utilizes gradient-elution HPLC with an applied voltage.

The practicality of these approaches varies for biopolymers. This discussion covers true biopolymers, including carbohydrates (polysaccharides); nucleic acids and oligonucleotides (oligos); and proteins, peptides, and antibodies. Not covered are modified nucleosides, amino acids, mono- and disaccharides, lipids (fatty acid esters), and smaller monomers leading to biopolymers.

6.3 Carbohydrates

Palm and Novotny have described the generation of a monolithic capillary bed formed by copolymerization of polyacrylamide and poly(ethylene glycol) with added ratios of acrylic or vinyl sulfonic acid (to produce the desired EOF) [41]. Hydrophobic ligands, including C_4, C_6, and C_{12}, were introduced via various acrylate esters and copolymerized with the monomers, leading to the basic backbone of a mixed polyacrylamide/poly(ethylene glycol) phase. Various ratios of these monomers were polymerized in a capillary whose walls were coated with and activated by a bifunctional reagent (such as an unsaturated, trimethoxysilane). The presence of free double bonds on the walls permitted the covalent attachment of the monolithic polymer gel. Various ratios of linear and cross-linked monomers (%T/%C) were used to generate different gel matrices with free sulfonic acid groups and were utilized for the separations of peptides and carbohydrates, as well as smaller organic molecules.

In order to visualize carbohydrates, which are usually UV transparent, the authors utilized a now-standard derivatization method involving reductive amination with 2-aminobenzamide and laser-induced fluorescence (LIF) detection with a He-Cd laser (325 nm) as the excitation source [42]. These derivatization-detection approaches have long been utilized in capillary LC (CLC) and HPCE modes. Precapillary tagging usually leaves an excess of the reagent in the final injection solution, not normally a problem if the reagent is not FL active at the ex/em wavelengths optimal for the derivatives. However, the excess reagent can interfere in the final chromatogram if FL active, or it can alter the surface activity of the stationary phase and adversely affect the reproducibility of the analytical procedures. An altered surface activity, in general, is detrimental to CEC reproducibility.

These studies involved isocratic CEC, without any external pressure-driven flow, and without step-gradient conditions. Detection was off-gel (after the gel portion of the packed capillary) using a fiber-optic cable placed as near as possible to the interface between the gel and free solution. Under such separation–detection conditions, the efficiencies for the Glc1-Glc3 oligos were between 190,000 and 230,000 theoretical plates per meter.

Figure 1 illustrates a typical isocratic chromatogram of a maltose–oligosaccharide mixture, where the reagent peak appears between 14 and 16 min (Figure 1B) [41]. Figure 1A is an enlargement of Figure 1B, which illustrates the elution of the excess tagging reagent after elution of the derivatized maltooligosaccharides (Glc7 to Glc1). Carbohydrate or protein separations with this particular packing/phase are pH and organic content dependent, conditions that greatly affect resolution and peak shape as well as total analysis time [41]. Figure 2 further illustrates the analysis of Glc4-Glc10 oligosaccharides, this time on a longer capillary and at a lower field strength. The larger oligomers (presumably Glc11-Glc16) were also visible, though poorly resolved from one another. The peak shapes, resolutions, efficiencies, and total analysis times, were also a function of the nature of the hydrophobic

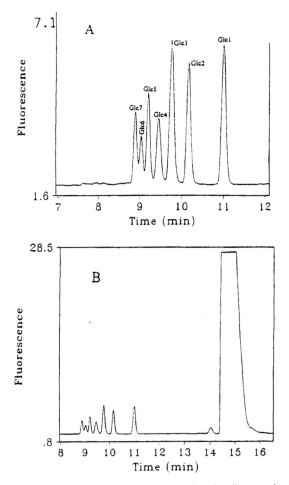

Figure 1. (A) Isocratic electrochromatogram of maltooligosaccharides (glucose (Glc1)-maltohexaose (Glc6)) in a capillary filled with a macroporous polyacrylamide/poly(ethylene glycol) matrix, derivatized with a C_4 ligand (15%) and containing vinylsulfonic acid (10%). 2-Aminobenzamide was used to tag the oligosaccharides for the laser-induced fluorescence detection. (B) is the same analysis as in (A), including the peak of the derivatization agent, which appears at 14–16 min. Conditions: capillary, 32 cm (25 cm effective length) × 0.1 mm i.d.; field strength 900 V/cm, 20 µA (current); sample concentration, 5–10 µM; detection, UV absorbance at 200 nm; temperature, 22°C; mobile phase (buffer), 10 mM Tris/15 mM boric acid (pH 8.2); other conditions are described in Ref. [41]. (From Ref. [41], with permission.)

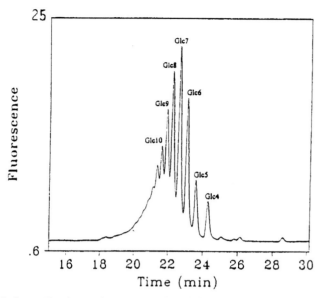

Figure 2. Isocratic electrochromatography of the oligosaccharide ladder in a capillary filled with a macroporous polyacrylamide/poly(ethylene glycol) matrix, derivatized with a C_4 ligand (15%) and containing vinylsulfonic acid (10%). Conditions: capillary length, 50 cm (40 cm effective length) × 0.1 mm i.d.; mobile phase, acetic acid 1:1000 containing 5% (v/v) acetonitrile; field strength, 600 V/cm, 14 μA; injection, 5 s (100 V/cm); sample concentration, 30 mg/mL in derivatization solvent and thereafter diluted 1:100 in the mobile phase; detection, UV absorbance at 200 nm; temperature, 22°C; mobile phase (buffer), 10 mM Tris/15 mM boric acid (pH 8.2); other conditions are described in Ref. [41]. (From Ref. [41], with permission.)

moieties incorporated into the final, monolithic polymer gels (C_4, C_6, C_{12}, or others possible).

The advantages of using a monolithic gel polymerized in situ stationary phase, as in CGE, are several-fold. Good separations can be achieved, at times in less than 10 min, and often in less than 5 min. The migration time reproducibility is better than 1% (RSD) from run to run and 2.5% from day to day. Finally, the gel is stable up to at least 50% acetonitrile when used in the mobile phase [41].

Additional work in the area of carbohydrate analysis by CEC has been reported by El Rassi and colleagues [43–46]. In their approach toward improved separations of carbohydrates, they designed a special octadecyl–silica (ODS) stationary phase for CEC that had a limited amount of hydrocarbon coverage in order to leave 75% of the surface silanols unreacted. This yielded a relatively moderate EOF, yet exhibited RP behavior toward alkylbenzene homo-logs and a series of p-nitrophenylglycosides and p-nitro-phenyl-maltooligosaccharides, all within a relatively short period of time [44].

A prerequisite for achieving the separation and detection of carbohydrates by CEC (or capillary zone electrophoresis [CZE] or CLC) with ODS capillary columns is to derivatize the sugar analytes with fluorophores (as above) or chromophores to yield preferably neutral derivatives, though at times charged derivatives can be utilized. In these studies, the p-nitrophenyl group was introduced at the terminal end of the carbohydrates, singly tagged. The rationale for carbohydrate derivatization, then, is really twofold: (1) to increase the sensitivity of the detection and (2) to confer the hydrophobicity necessary for RP-CEC. Under these CEC conditions, even alpha- and beta-anomers of some p-nitrophenyl-monosaccharides were readily separated in the presence of a small amount of borate buffer in the hydroorganic eluent in CEC. Such conditions have also been utilized by these and other workers in the past using CZE/CE [44].

Figure 3 shows the CEC electrochromatograms of p-nitrophenyl-α-D-glucopyranoside and p-nitrophenyl-α-D-maltooligosaccharides. The percentage of ACN (v/v) in the mobile phase was changed in order to determine the optimal mobile phase composition for rapid elution time and high separation efficiency. These studies were done in a true isocratic CEC mode, no pressure-driven flow was used, and the capillaries were packed using a wet slurry packing approach, with wet bare silica of 5 μm average particle size being sintered at the ends by a Bunsen burner to form the end frits. Detection was on-column, through the packed bed, rather than off-bed, as above [41]. Satisfactory separa-

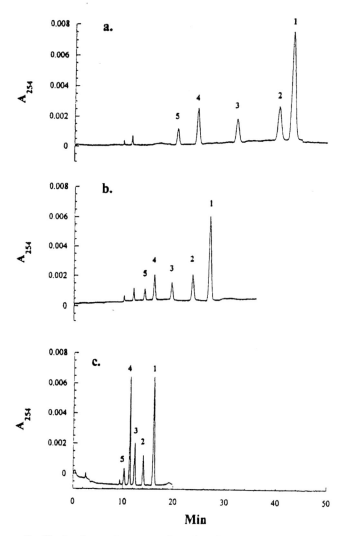

Figure 3. Electrochromatograms of p-nitrophenyl-α-D-glucopyrano-sides and maltooligosaccharides. Mobile phase: (c), 40% v/v of 5 mM NaH_2PO_4 (pH 6.0), 40% v/v H_2O and 20% v/v acetonitrile; (b) 42. 5% v/v of 5 mM NaH_2PO_4 (pH 6.0), 42. 5% v/v H_2O and 15% v/v acetonitrile; (a) 45% v/v of 5 mM NaH$_2$PO4 (pH 6.0), 45% v/v H_2O and 10% v/v acetonitrile; voltage, 10 kV; capillary column, 20 (27) cm × 100 μm packed with OAS under 3000 psi for 30 min; voltage, 20 kV; detection wavelength, 254 nm; 18°C; other conditions are described in the original publication [43]. Solutes: 1, p-nitrophenyl-α-D-glucopyranoside; 2, p-nitrophenyl-α-D-maltoside; 3, p-nitrophenyl-α-D-maltotrioside; 4, p-nitrophenyl-α-D-maltrotetraoside; 5,

155

tion was obtained with a mobile phase of low acetonitrile content (20% v/v) and low electric field strength (370 V/cm).

In terms of retention time, the reproducibility of the CEC system was quite good, with %RSD <0.55. The column separation efficiency was high, varying between 72,000 and 152,000 plates per meter for the various derivatives. The elution order of typical p-nitrophenyl saccharides was the same as that observed in RPC and is believed to be due to the hydrophobicity of the glycosides as well as to organic-induced conformational changes of the glycosides [45]. Several other examples of CEC separations of saccharides are offered in this study, including the separation of various α- and β-anomers of glucopyranoside derivatives. Table 2 summarizes the realized column efficiencies in terms of plates per meter and retention time reproducibility (%RSD) for a number of these carbohydrate derivatives in this type of packed bed CEC [43].

Table 2. Column Efficiency and Retention Time Reproducibility[a]

p-Nitrophenyl derivative of	Retention time (%RSD)	Column efficiency (plates/m)
Galactose	0.54	95,000
Glucose	0.47	92,000
N-acetylglucosamine	0.21	83,000
Mannose	0.31	84,000
Maltose	0.55	95,000
Maltotrioside	0.23	85,000
Maltotetraoside	0.49	152,000
Maltopentaoside	0.26	72,000

[a]Mobile phase: 20:80% v/v acetonitrile, 3.34 mM sodium phosphate, pH 6.0; capillary column: 20 cm (27 cm total length) × 100 μm i.d. packed with 5 μm ODS; 10 kV; λ = 254 nm; pressure injection at 20 psi for 10 s.
Source: From Ref. [44], with permission

6.4 Peptides, Proteins, and Antibodies

CEC has shown utility in the separation of the enantiomers of amino acids [51–54]. Most of the separation methods utilized molecularly imprinted polymers with recognition sites as the stationary phase. Other groups have successfully utilized gradient elution to separate mixtures of dansylated amino acid mixtures on an ODS (octadecylsilane) stationary phase (pp. 126–128) [54].

To simplify the discussion, this chapter treats the general class of biopolymers (peptides, proteins, and antibodies) as peptides, realizing that proteins are really just larger peptides of higher molecular weight. Antibodies are really just larger proteins of specific conformations, shape, size, and immunogenicity together with antigenic recognition properties [47–50]. The various modes of CEC can be applied to peptides of varying size and complexity.

Palm and Novotny applied the polymeric gel beds (monoliths), described previously for the separation of carbohydrates, for peptide resolutions in CEC [41]. The peptide separations used the same packing material, but instead with 29% C_{12} as the ligand. Additional CEC conditions are indicated in Figure 4, which depicts the separation of a series of tyrosine-containing peptides, with detection at 270–280 nm. In this particular study, peptide elution patterns were very sensitive to changes in pH and ACN concentrations. A gradient-elution technique, not employed here, would have been more appropriate for such samples of peptides having small differences in their constitution. Attempts to elute protein samples were unsuccessful with these particular gel matrices, perhaps due to the high hydrophobicity of the packings [41].

Euerby et al. reported the separation of an N-methylated C- and N-protected tetrapeptide from its non-methylated analog (Figure 5) [55]. These separations utilized a Spherisorb ODS1, 3-μm packing material, without pressure-driven flow (true CEC using a commercially available CE instrument), and an ACN/buffer. Using non-optimized, non-pressurized CEC conditions (non-PEC, non-pressure driven CEC),

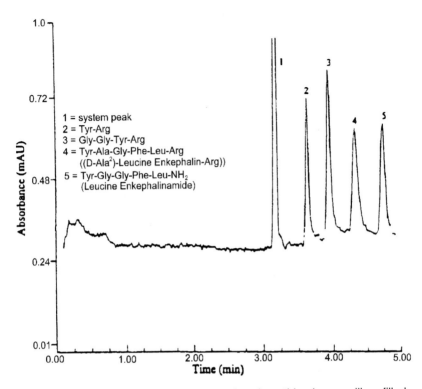

Figure 4. Isocratic electrochromatography of peptides in a capillary filled with a macroporous polyacrylamide/poly(ethylene glycol) matrix, derivatized with a C_{12} ligand (29%) and containing acrylic acid. Conditions: mobile phase, 47% acetonitrile in a buffer; voltage, 22.5 kV (900 V/cm), 7 μA; sample concentration, 4–10 mg/mL; detection, UV absorbance at 270 nm; other conditions are described in Ref. [41]. (From Ref. [41], with permission.)

separation of the two tetrapeptides could be achieved in a run time of 21 minutes, with efficiency values of 124,000 and 131,000 plates per meter.

In comparison, when a pressurized CE system was used (buffer reservoirs and capillary were pressurized, without pressure-driven flow of buffer), separation of the components was achieved within 3.5

minutes (Figures 5C and 5D). According to Euerby et al., the separation of these two peptides using a pressure-driven HPLC gradient analysis took 30 minutes and gave comparable peak area results.

Although this study illustrated the improved efficiency of both nonpressurized CEC and pressurized CEC over HPLC, the conclusions cannot be generalized. The scope of this peptide study was limited to two tetrapeptides of unspecified structures, distinguishable by a methylated nitrogen group. The low voltage used in the study produced a low EOF that allowed sufficient time (21 minutes) for chromatographic separation without pressurization of the vials. The difference between the CEC elution time of 21 minutes and the HPLC elution time of 30 minutes are not as convincing as Euerby states. Furthermore, this study does not offer any insight into the capabilities of CEC with diastereomic and enantiomeric peptides. Using recognition sites on molecularly imprinted polymers with predetermined selectivity, nondansylated D,L-leucine could not be separated [56]. The study by Euerby et al. does not offer any insight into the simultaneous separation of numerous peptides, as in a complex peptide map from a large peptide or antibody.

While chemometric optimization of operating conditions has been done for many years in GC, HPLC, and CE, the literature is lacking in true chemometric software approaches to CEC methods optimization [57–59]. It is not clear what, if any, analytical figures of merit have been used to define an optimized set of conditions for biopolymer analyses by CEC or why specific stationary phases (packings) have been selected as the optimal support in CEC applications for biopolymers. It is hoped that more sophisticated optimization routines, especially computerized chemometrics (expert systems, theoretical software, or simplex/optiplex routines), will be employed from start to finish.

The coupling of an ESI mass spectrometer with a PEC system has been shown to separate peptides [60]. Schmeer et al. utilized a commercial packing material, namely a reversed-phase silica gel, Gromsil ODS-2, dp = 1.5 μm, already utilized in capillary HPLC for peptide

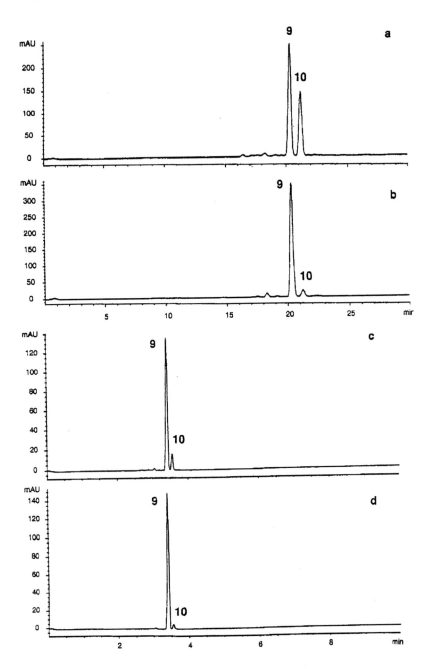

separations. It is possible that the EOF alone with this packing was insufficient to elute all peptides in a reasonable time frame, and thus pressurized flow was introduced. PEC also provides a more stable EOF and prevents bubble formation. No gradient elution PEC was demonstrated in this particular study. A mixture of enkephalin methyl ester and enkephalin amide were separated using the packed capillary column (Figure 6). The coupling of these two methods showed enhanced sensitivity and detectability (LODs) at low concentrations (20 ppm injected, 3 pmol detected at MS).

As in the Euerby study, the one by Schmeer et al. offers little insight into the true capabilities of CEC to separate peptides, as in a complex mixture of peptides [60]. Enkephalin methyl ester and enkephalin amide contain significant physical differences, which influenced their elution times. As one would expect, the mass (electro)pherogram reveals that the enkephalin amide eluted almost 2 min faster than the enkephalin methyl ester. Since the acetonitrile/water/tri-

---

Figure 5. Separation of synthetic, protected tetrapeptide intermediates. Peak assignments: (9) *N*-methyl C- and N-protected tetrapeptide; (10) non-*N*-methyl C- and N-protected tetrapeptide. The structures of these compounds is proprietary and was therefore not disclosed. Detection wavelength: 210 nm with a 10-nm bandwidth and a 1-s rise time. Electrochromatography was performed on a 250-mm x 50-μm i.d. 3-μm Spherisorb ODS1 packed capillary using an acetonitrile/Tris (50 mmol/L, pH 7.8) buffer 80:20 v/v mobile phase, capillary temperature of 15°C, and an electrokinetic injection of 5 kV/15 s). (a) Synthetic mixture of protected tetrapeptides 9 and 10. Efficiency values of 124,000 and 131,000 plates per meter were obtained for analytes 9 and 10, respectively. Unpressurized HP3D CE system using an applied voltage of 5 kV. (b) Chromatogram of synthetically prepared 9, the presence of residual nonmethylated tetrapeptide (10) can be clearly seen. Unpressurized HP3D CE system using an applied voltage of 5 kV. (c) Chromatogram of synthetically prepared 9, spiked with 10% of the nonmethylated tetrapeptide (10). Efficiency values of 83,000 and 101,000 plates per meter were obtained for analytes 9 and 10, respectively. Pressurized CE system using an applied voltage of 30 kV. (d) Chromatogram of synthetically prepared 9, the presence of residual nonmethylated tetrapeptide (10) can be clearly seen at the 3% level. Pressurized CE system using an applied voltage of 30 kV. Additional conditions are indicated in Ref. [55]. (From Ref. [55], with permission.)

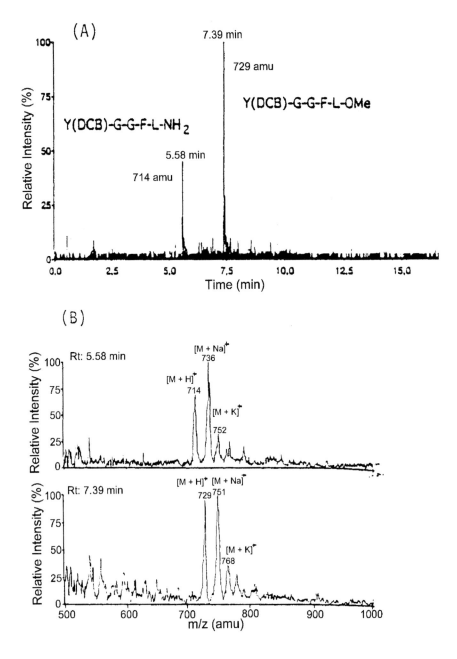

fluoroacetic acid buffer solution was at a low pH, one would expect a distinct separation based on the electrophoretic separation of these two peptides. The amide should migrate faster to the anode than the ester since it carries an additional positive charge under acidic conditions. In addition, since these two compounds are structurally distinctive, a chromatographic separation was probably also taking place.

The study does provide a nice example of a peptide separation based on chromatographic and electrophoretic separation mechanisms probably occurring simultaneously. It also demonstrates how easily CEC and PEC interface with ESI-MS.

Several studies by Lubman's group have used OT or packed-bed capillaries in CEC, at times with external pressure-driven flow to generate PEC [61–64]. In this work, both OT-CEC as well as packed CEC approaches were interfaced with an on-line ion trap storage/reflectron time-of-flight mass spectrometer (ITS-TOFMS) [61].

Wu et al. reported the separation of a six-peptide mixture using an open tubular column (OTC) (C-8 coating) CEC coupled to an on-line ITS-TOFMS [61]. Their work included an interesting experimental trick to enhance the EOF under acidic conditions without using pressure-driven buffer flow. After preparing the CEC column with a commercial C-8 stationary phase, they chemically coated the inner surface of the capillary wall with an amine, (3-aminopropyl) trimethoxysilane (APS). This coating served two purposes: First, it significantly enhanced the EOF in acidic buffer solutions, so that a large EOF flow rate was obtained without using a very high voltage, which in turn resulted in a stable flow. Second, the surface silanol groups were covered by the amine groups, which carried positive charges in an acidic solution, so that nonspecific adsorption between

---

Figure 6. Interfacing of pressure driven CEC (PEC) for the separation of two simple peptides, enkephalin methyl ester (5.58 min) and enkephalin amide (7.39 min). (A) Extracted mass chromatogram of m/z 714 and 729 for the on-line peptide separation. (B) Mass spectra taken from the chromatographic peaks in (A), illustrating true Mr, and the presence of M+ H, M + Na, and M + K cations at appropriate m/z (amu) values. Specific operating conditions are indicated in Ref. [60]. (From Ref. [60], with permission.)

the peptide sample and the inner surface was greatly reduced. Because of the high flow velocity, a six-peptide mixture could be separated to baseline within 3 min on this system, with absolute identification of each peptide peak by the ITS-TOFMS combination. With a cationic coating on the capillary wall, the EOF in the capillary is reversed. Because of the high duty cycle of the MS and the column path length–independent concentration-sensitive feature of the ESI process, high-quality total ion chromatograms (TICs) could be obtained with injections of only 1–2 fmol of peptide samples. A concentration limit of detection of 1×10^{-6} M was also achieved due to the preconcentration ability of CEC. Figure 7 shows the separation of the six-peptide mixture on an APS-coated OT-CEC column. It was also possible to obtain complete ESI mass spectra (multiply charged Mr ions) for each of the peaks in Figure 7.

Wu et al. also demonstrated the ability of interfacing true gradient elution CEC with OTC to the same ITS-TOFMS detector with ESI [61]. The separation was performed in true CEC fashion, without any external pressure-driven mobile phase flow. Using a very simple gradient formation device with a single syringe pump and a anodic buffer vial, constant, reproducible changes in buffer composition, and therefore a true gradient, was observed. Figure 8 illustrates both the UV and TIC traces of a gradient CEC separation of the tryptic horse heart myoglobin digest, illustrating 10 peaks almost baseline resolved within 6 min. Among the peaks shown in the TIC, 15 usable mass spectra (including coeluting components) could be obtained to cover about 90% of the amino acid residues in the protein. The calculated and measured masses were found to be in excellent agreement, as presented in Table 3. These mass spectra were all obtained with high-resolution conditions (~1500), even though they were all obtained at a full mass range sampling speed of 8 Hz. Clearly, another advantage of using the MS as an on-line detector for CEC—or for HPLC or CE—is that it can identify some partially resolved or unresolved peaks, thereby increasing the resolving power of the basic CEC system. This has also been amply demonstrated

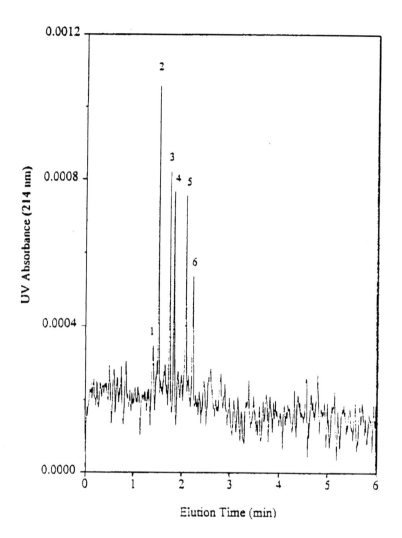

Figure 7. Open tubular CCE separation of a six-peptide mixture using a column with APS coating. Separation conditions: column length, 30 cm (25 cm to detector) × 9 μm (i.d.) × 150 μm (o.d.); separation voltage, −12 kV; injection, −2 kV × 3 s; sample concentration, 1 x 10^{-5} M; UV detection at 214 nm. The six peptides are: (1) methionine enkephalin, (2) bradkykinin, (3) angiotensin III, (4) methionine enkephalin-Arg-Phe, (5) substance P, and (6) neurotensin. Additional conditions are indicated in Ref. [61]. (From Ref. [61], with permission.)

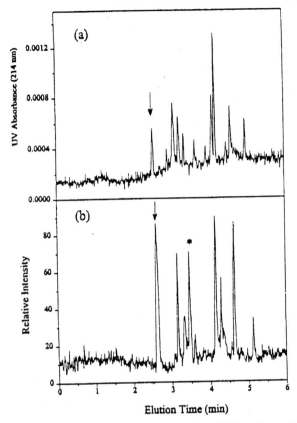

Figure 8. Protein digest using gradient-elution CEC with on-line ITS-TOFMS. UV trace (a) and TIC (b) of gradient CEC separation of a tryptic horse heart myoglobin digest. Conditions: 0–35% acetonitrile gradient in 6 min; column length, 40 cm (for UV detection, 35 cm to UV detector) × 9 μm (i.d) × 150 μm (o.d.); separation voltage, –14 kV; injection, –2 kV x 5 s; UV detection at 214 nm; MS detection speed, 8 Hz. Additional conditions are indicated in Ref. [61]. (From Ref. [61], with permission.)

for HPLC-MS and CE-MS approaches with peptide maps and intact proteins [74–77].

There are several other articles by the Lubman group that deal with peptide analysis using various forms of CEC interfaced to ITS-TOFMS, again using ESI as the interface [62–64]. Much of this work

Table 3. Comparison of Calculated and Measured Tryptic Fragments of Horse Heart Myoglobin from CEC-MS Analysis

No.	Fragment	Calculated mass (Da)	Measured mass (Da)	Sequence
1	1–16	1816.0	1816.4	GLSDGEWQQVLNVWGK
2	17–31	1606.8	1606.3	VEADIAGHGQEVLIR
3	32–42	1271.4	1271.4	LFTGHPETLEK
6	48–50	396.5	396.4	HIK
7	51–56	707.8	707.6	TEAEMK
8	57–62	661.7	661.8	ASEDIK
10	64–77	1378.7	1379.0	HGTVVLTALGGILK
13	80–96	1854.1	1853.6	GHHEAELKPLAQSHATK
14,15	97–102	752.9	752.7	HKIPIK
15	99–102	469.6	469.4	IPIK
16	103–118	1885.2	1885.5	YLEFISDAIIHVLHSK
17	119–133	1501.6	1501.2	HPGNFGADAQGAMTK
18	134–139	747.9	748.1	ALELFR
20	146–147	309.4	309.5	YK
21	148–153	649.7	649.6	ELGFQG

aAverage mass of all charge states of the fragment observed.
From Ref. [61], with permission.

deals with the separation of tryptic digests of proteins using various buffers in pressurized CEC or PEC, again coupled to IT-TOFMS [63–64]. The work utilized packed-bed CEC, often a Vydac C18 silica gel phase (3 mm), as with peptide separations used in HPLC. PEC was utilized for the analysis of peptide mixtures and protein digests. The PEC mode was used because these phases yield very low EOF at low pH. By using both low EOF and pressure-driven flow, as opposed to

EOF alone, the total time for analysis of peptide mixtures could be reduced. Efficiency and plate counts were less than would have accrued by isocratic or gradient CEC alone.

Gradient-elution packed-bed PEC was utilized with on-line ITS-TOFMS in order to demonstrate the advantages of combining these two separation schemes. The supplementary pressure-driven flow suppressed bubble formation and allowed for the tuning of the elution of peptides using the applied electrical field. It also stabilized the overall EOF. In this way, a very fast separation of six peptides could be performed. Using short, 6-cm columns, a tryptic digest of bovine cytochrome c was fully separated in about 14 min by properly tuning the applied voltage and the external pressure from an on-line HPLC pump. Fairly complex protein digests, such as that from chicken oval-bumin, containing more than 20 peaks, could be resolved in the TIC in 17 min. Again, the use of an on-line ITS-TOFMS detector increased the resolving power of the PEC system, by providing for absolute identification of coeluting components. Sample concentrations were in the range of 5×10^{-6} M, with an injection volume of 1.5 mL [62–64].

In another review, Lubman et al. point out that there are some major advantages in using OTC for CEC compared with packed-bed CEC [62]. OTCs with inner diameters of around 10 µm have been found to have a smaller plate height compared with that for packed columns of the same internal diameter. It is also possible that some packed columns may have smaller HETP than OTC-CEC capillaries. This is due to the lack of band-broadening effects associated with the presence of packing particles and end-column frits. OTC capillaries do not require end frits. High concentration sensitivity is another advantage of OTCs, since columns with very small dimensions are used. The small diameter of the OTCs allows for the use of a higher field strength in CEC, without significant Joule heating. OTCs can also often provide more rapid separations than packed columns, by eliminating intraparticle diffusion, which is the dominant limitation for ultrafast separations in packed columns.

However, there are some grave difficulties involved in using OTCs, perhaps because of the real difficulties with sample injection and detection. The injection volume of OTCs is in the low nL or even pL range. The very small inner diameters of most OTCs make optical detection difficult, but they are clearly very compatible with a concentration sensitive detection method, such as ESI-IT–TOFMS. This approach is independent of the optical path length of the capillary, and thus the major disadvantages of OTCs may be overcome in a CEC-ESI-MS approach and configuration. With peptide mixtures, gradient-elution CEC, with or without pressure-driven flow, is almost required over isocratic or step-gradient methods, since small changes in the mobile phase composition result in large changes in peptide retention times. It is thus difficult to separate a complex peptide mixture, such as a large protein digest, using the isocratic mode in CEC or HPLC or CLC.

It should be noted that Lubman's studies used pressurized flow together with applied voltages [61–64]. For very complex peptide mixtures, it may prove difficult to optimize the selectivity of peptides using just the applied field in CEC without any supplementary pressure. If one attempts to improve selectivity by altering pH or ionic strength, then the EOF may be adversely affected and the total analysis times may be impractically prolonged. On the other hand, by applying the external pressure-driven flow, total analysis time can be reduced, the effect of applied voltage on peptide selectivity can be optimized independent of external flow rate, and selectivity can be obtained without giving up shorter analysis times. Thus, the tuning of the selectivity of peptides becomes possible, since the applied voltage and the supplementary pressure are available as two totally separate, tunable parameters that can be optimized for achieving a higher selectivity than would be possible using either HPLC or CE modes alone.

This point further illustrates the advantages of using combined pressure-driven flow with an external applied (variable) voltage and gradient-elution, RP, packed-bed CEC conditions. Wu et al. show in Figure 9 the TICs of the separation of a bovine cytochrome c digest, using a 6-cm-long column with gradient elution and a packed-bed CEC capillary with a commercial C_{18} packing [64]. In Figure 9a, no separation voltage was applied, resulting in a conventional gradient RP CLC separation of a peptide mixture (90 bar). The use of such a short column in HPLC made it difficult to resolve all the components in the digest. The peaks marked by asterisks contain two eluting components. In Figure 9b, 2000 V was applied to the column, and the back pressure was reduced to 50 bar. It is often advantageous when using PEC to balance the EOF being controlled by the applied voltage with the applied pressure flow in order to maintain a relatively constant mobile phase flow rate, since changes in flow rate can affect the separation.

In Figure 9b, all of the peptides eluted faster than in Figure 9a, because of the EOF, EPF migrations, and applied voltage. The peaks in Figure 9b are all sharper, indicating an increase in separation efficiency. Certain peaks that were unresolved in Figure 9a (marked by an arrow) are resolved in Figure 9b, mainly because of the EPF separation of these two components. However, the first two peaks in Figure 9a are coeluted in Figure 9b, since RP partition mechanisms cause the EPF to be in the opposite direction of the separation for these two fragments. Figure 9c illustrates the overall effect of a larger applied voltage (1400 V), but the peak marked by the asterisk in Figure 9c remains unresolved. In Figure 9d, the applied voltage was changed to 600 V, with a supplementary pressure of 70 bar. This combination of pressure and voltage resulted in resolution of the four previously unresolved peaks [64].

The report by Wu et al. is an ideal illustration of utilizing all possible variable parameters in PEC: the nature of the gradient, the nature of the packing, the dimensions of the packed capillary, the applied voltage, the nature of the mobile phase (pH, ionic strength, organic/aqueous), and the applied pressure-driven flow component [28]. As Figure 9 demonstrates, by varying both applied pressure (bar) and applied voltage (V), optimization can be achieved without varying the gradient conditions or the packed bed. The use of external applied

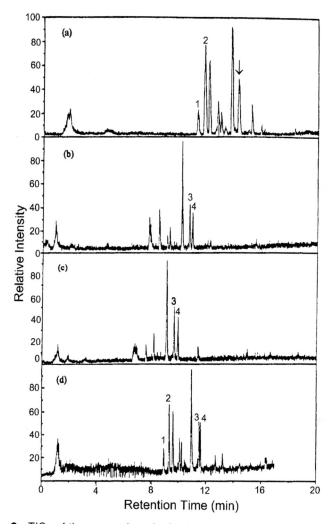

Figure 9. TICs of the separation of a bovine cytochrome c digest using a 20 min, 0–50% acetonitrile gradient with a packed Vydac C-18 (3-μm) bed, with sample injections of 8 pmol, corresponding to the original protein. Column length, 6 cm × 180 μm i.d. × 360 μm o.d.; column operating conditions: (a) HPLC mode with a back pressure of 90 bar; (b) 1000 V applied voltage with 50 bar supplementary pressure; (c) 1400 V applied voltage with 50 bar supplementary pressure; and (d) 600 V applied voltage with 70 bar supplementary pressure. Additional conditions are indicated in Ref. [64]. (From Ref. [64], with permission.)

pressure (flow rate) together with varying voltages appears to provide additional analyte selectivity not possible in simple isocratic/gradient CEC, perhaps not possible in isocratic/gradient CLC, and surely not possible in CE without any packing present. This approach also provides selectivity on the basis of partitioning mechanisms (nature of the RP stationary phase) as well as electrophoretic differences, with the latter utilized in CZE.

In conventional CLC or HPLC, there is no applied voltage to introduce EPF differences to improve selectivity performance. This distinction, especially when combined with on-line MS, underscores the greatest advantage of gradient-elution PEC. The only drawbacks, compared with pure CEC, are lower plate counts and reduced column efficiency resulting from application of an external pressure [64].

In contrast to the mainly packed-column methods described above, Pesek et al. utilized the OTC approach for CEC separation of peptides and proteins [65–68] (see also discussion in Section 2.6, p, 53). As mentioned above, OTC has several real advantages over packed-bed capillaries in CEC or CLC. Pesek has engineered certain novel approaches to bonding of organic ligands to silica surfaces and has demonstrated the increased stability of such an approach to covalent silicon bonding in HPLC, HPCE, and CEC studies [65–68].

A wide variety of bonded phases and analyte applications have been reported. While conventional attachment of organic ligands to silica has usually involved a siloxane (Si-O-Si) bond or variations thereof, Pesek's approach realizes a much more pH- and aqueous-stable Si-C bond. In this approach, the fused silica inner surface was etched with ammonium hydrogen difluoride to increase the surface area. The etching produced radial extensions of the ligand from the surface to facilitate solute-bonded phase interactions. In theory, the etching process increased the surface area of the inner walls sufficiently to induce solute interactions with the capillaries modified with an organic moiety. The chemistry used to modify the etched capillary was based on a silation–hydrosilation reaction scheme, which leads to a direct silicon–carbon bond on the surface:

$$\geqslant Si - OH + (OEt)_3 SiH \longrightarrow \; \geqslant Si - O - Si - H + 3\; EtOH$$

$$\geqslant SiO - Si - H + R - CH = CH_2 \longrightarrow \geqslant Si - O - Si - CH_2 - CH_2 - R$$

In this process, the etched surface of the capillary was first reacted with triethoxysilane (TES) to produce a hydride surface. An organic moiety was then attached to the hydride intermediate by passing a solution containing a terminal olefin and a suitable catalyst, such as hexachloroplatinic acid, through the capillary. In order to characterize the behavior of this new electrochromatography format, Pesek and colleagues ran a test mixture of five polypeptides and proteins on four types of capillaries and obtained the results summarized in Table 4 [67].

As one would expect, in the bare capillary there was a relatively small difference between the migration times of the five components. The etched capillary also had a relatively small range of migration times, but each component had a much lower linear velocity than in the

Table 4. Migration Times of Varioous Prptides and Proteins

	Capillary and buffer				
Compound	Bare pH 3.7	Etched (300°C, 3 h): pH 3.7	Si–H modified: pH 3.7	C_{18} modified: pH 3.0	C_{18} modified: pH 3.0 with 10% MeOH
Lysozyme (turkey)	1.83	Wide peak	2.23	4.06	4.67
Angiotensin III	2.05	6.82	3.13	4.68	5.37
Bradykinin	2.09	6.98	3.00	6.01	5.68
Ribonuclease A	2.02	6.89	2.89	6.95	6.47

From Ref. [67], with perrmission.

bare capillary. Even though the effective length of the etched capillary was twice that of the bare capillary, the migration time was more than twice as long. This was likely due to the fact that the etching process reduced the number of free silanols on the surface and hence lowered the EOF. Some improvements in separation as well as an increase of migration time over the bare capillary were seen for the hydride-modified column. The increase in elution time was probably due to an increase of free silanols, which led to a higher EOF. The relatively close elution of the solutes in the hydride capillary indicated that their separation was based mainly on EPF mobility and not on interactions with the capillary surface. When the same solutes were evaluated using the C_{18}-modified, etched capillary, some solutes had long retention times and/or poor peak shapes.

Since the result of the hydrosilation reaction was to replace hydrides with octadecyl moieties, there should have been no increase or a decrease in the number of silanols on the surface. Therefore, the change in times reported in Table 4 must have been due to increased interaction with the newly modified surface. The increase in peak width also supported this conclusion, since there should be some decrease in efficiency due to mass transfer effects. Pesek et al. attributed the long retention times and poor peak shapes to the unfolding of the proteins and polypeptides, which was a result of the strong interactions between the solutes and the bonded octadecyl moiety. To alleviate these problems, they lowered the pH from 3.7 to 3.0. This increased the charge on the species and caused each solute to elute faster and with better peak symmetry (Figure 10) [67]. The increased peak widths and the larger range of elution times indicated that chromatographic interactions between the solutes and the bonded moiety occurred, as opposed to pure electrophoresis.

Further support for this conclusion was obtained when methanol was added to the mobile phase, thus causing a decrease in elution times for some solutes and a slight improvement in efficiency, and indicating a decrease in capacity factor, k'. Interestingly, the first two components in the mixture showed an increase in elution time when methanol was

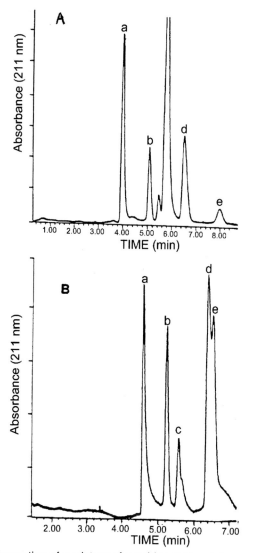

Figure 10. Separation of a mixture of peptides and proteins on a C_{18}-modified etched capillary at pH 3.0 with (A) 0% methanol (I = 18 mA) and (B) 10% methanol (I = 14 μA) in the electrolytes. Conditions: total capillary length = 45 cm; effective length = 25 cm × 50 μm (i.d.) × 575 μm (o.d.); buffer = 50 mM citric acid and 245 mM β-alanide; V = 25 kV; solutes: a = lysozyme (turkey); b = angiotensin III; c = bradykinin; d = ribonuclease A; and e = angiotensin I. (From Ref. [67], with permission.)

added to the electrolyte (Figure 10). This indicated that electrophoretic as well as bonded-phase interactions were contributing to the separation.

An additional experiment performed by Pesek et al. showed the stability and reproducibility of their capillaries. Thirty-one consecutive injections of lysozyme were performed, followed by an identical series of 31 injections of ribonuclease A. No discernible increase or decrease in retention time for either protein was observed. The reproducibility of each result was ±1.5% [67].

In a later study, Pesek et al. reported the separation of other proteins using a diol stationary phase [66]. The use of a diol stationary phase should result in a surface that is more hydrophilic than a typical alkyl bonded moiety, such as C_{18} or C_8. Figure 11 depicts the separation of basic proteins in a buffer with a pH value of 4.41. The peaks are relatively symmetric, indicating that little adsorption of the solutes on the etched and modified surface took place. Although no mention was made of the small, broader peaks near the baseline, they could have been due to partial unfolding of some of these proteins. A buffer of lower pH could alleviate this problem by affecting the charge of the protein species and their conformations.

A comparison of the separation characteristics for a series of angiotensins on bare, unetched, diol-modified etched, and C_{18}-modified etched capillaries is shown in Figure 12. For each column, the elution order was the same, indicating that while solute-bonded phase interactions may have been significant, the differences in electrophoretic mobility were primarily responsible for the separations observed with the angiotensins. The longer migration times on the C_{18} column could have been due to more efficient bonding (i.e., a greater number of bonded moieties per unit area) or to stronger solute-bonded phase interactions. The researchers attributed this greater retention time to the greater hydrophobic interactions between the bonded C_{18} and the solutes. The separation for the two modified, etched capillary columns was a combination of differences in EPF mobility, as well as solute-bonded phase interactions.

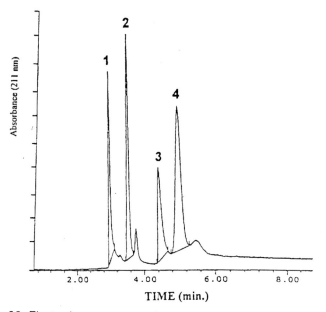

Figure 11. Electrochromatograms of protein separation on a diol capillary. total capillary length = 45 cm; effective length = 25 cm × 50 μm (i.d.) × 575 mm (o.d.); V = 22 kV; I = 7 μA; buffer = 25 mM β-alanide and 50 mM acetic acid; pH = 4.41; solutes: 1 = cytochrome c; 2 = lysozyme; 3 = myoglobin; and 4 = ribonuclease A. (From Ref. [66], with permission.)

The interesting results reported by Pesek et al. confirm the conclusion that proteins and peptides can be separated with chemically modified, etched fused-silica capillaries. The results showed that distinctive chemical modification (e.g., diol, C_{18}) of the packing yielded significant variations in retention times due to differences in solute-bonded phase interactions. Other factors, such as pH, can also influence this interaction, by affecting charge and protein conformations. Combining all these factors in the separation of peptides and proteins provides an experimenter with many important decisions to be made for the optimized experimental conditions to be used. Other chemical modifications of etched fused capillaries need to be studied in order to provide a better understanding of their interactions with proteins and peptides as well as other biopolymer classes.

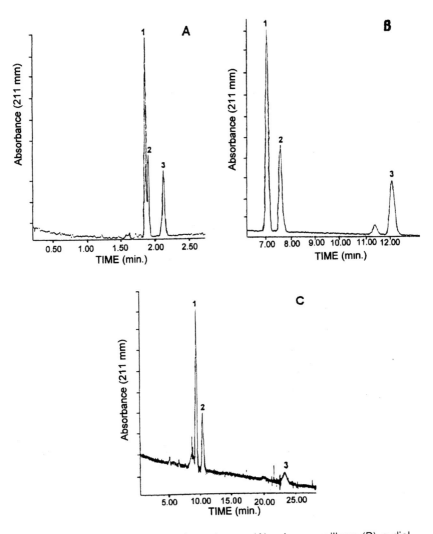

Figure 12. Separation of angiotensins on (A) a bare capillary; (B) a diol capillary; and (C) a C_{18} capillary; total capillary length = 45 cm; effective capillary length = 25 cm; V = 30 kV; I = 29 µA; pH = 2.14, solutes: 1 = angiotensin III; 2 = angiotensin I, and 3 = angiotensin II. (From Ref. [66], with permission.)

6.5 Nucleic Acids and Oligonucleotides

There has been much less work reported in CEC for this class of biopolymers than for others. Although there are several publications on the separation of monomeric nucleosides and nucleotides, they are not included here because they are not biopolymers [69–72].

One of the few reports on the separation of oligonucleotides is that of Behnke and Bayer, who used a pressurized, gradient-elution CEC apparatus [73]. In this study, termed pressurized gradient electro-HPLC (a variation of PEC, using an HPLC system with a voltage applied across the capillary length), the authors demonstrated improved separations (RP) of charged analytes (anionic) on a 5-μm C_{18} reversed-phase silica gel packing. The influence of applied voltage gradients up to 400 V/cm on the separation in both isocratic and gradient elution modes was studied. The authors also made a direct comparison among microbore HPLC, electro-HPLC, and MECC (also known as MEKC). As in some of the PEC applications of Lubman et al., applying a voltage to a micro-HPLC column results in EOF [61–64]. Its contribution to the overall velocity of the eluent increases with the electric field strength (V/L). Using this technique, analysis times are shortened and efficiency increases dramatically over micro-HPLC. Electric field strength, direction of the voltage, voltage gradient, pH, applied external pressure (bar), eluent composition, and overall mobile phase gradient are the most important other parameters that can be varied to optimize the performance of electro-HPLC.

Figure 13 shows the influence of applied voltage on the chromatographic separation of a series of oligonucleotides (dC_3–dC_{11}) and provides a comparison with MECC [73]. Higher order oligos have increasing capacity factors in RP-HPLC. The overall direction of the electric field can be used to retard (or advance) the elution of the higher homologues in order to optimize the selectivity of the separation. Separation may be improved by using pulsed voltages or pulsed-gradient voltages. Thus, in PEC or CEC, one can vary in a true gradient or step-gradient isocratic mode the effect of the stationary phase (C_1–C_{18}) as well as the effect of the direction and total applied

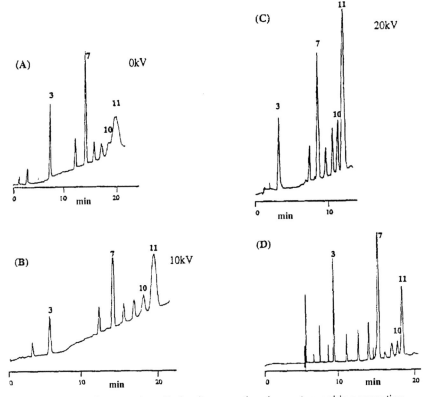

Figure 13. Influence of applied voltage on the chromatographic separation of oligonucleotides dC₃–dC₁₁ and a comparison with MECC. (A) gradient micro-HPLC; (B) electro-HPLC, applied voltage 10 kV; (C) electro-HPLC, applied voltage 20 kV; (D) MECC. Peaks 3, 7, 10 and 11 correspond to dC$_3$, dC$_7$, dC$_{10}$, and dC$_{11}$ oligonucleotides. Additional conditions are given in Ref. [73]. (From Ref. [73], with permission.)

voltage on the overall separations of anionically charged oligos. Figure 13A shows gradient micro-HPLC conditions, with no applied voltage, while Figure 13B shows true electro-HPLC, with an applied voltage of 10 kV. It is possible to vary the applied voltage or pressure-driven flow to any mixture of operational parameters and thus vary pressurized (parabolic) flow vs. EOF. The gradient electro-HPLC separations

(Figures 13B and 13C) for this oligonucleotide mixture of dC_3–dC_{11} demonstrated the optimization of efficiency and speed with increasing participation of EOF. Figure 13B shows an applied voltage of 10 kV, while Figure 13C shows a voltage of 20 kV; there was no applied voltage in Figure 13A.

Applying a voltage gradient drastically improved the efficiency of the gradient microbore separation of the oligos, as well as peak shape, efficiency, and overall resolution. These separations were compared to comparable separations using MEKC, without any packing material or external, pressurized flow, under isocratic conditions (normal CE). MEKC discriminates between analytes by a combination of both chromatographic and electrophoretic mechanisms. Both electro-HPLC and MECC showed excellent resolution and selectivity, above that which could be accomplished by just micro-HPLC means [73]. Electro-HPLC (Figure 13C) appears to offer everything that MEKC can provide in terms of analyte selectivity and resolution, as well as excellent peak shape and efficiency, but it can also greatly reduce the total analysis time, as illustrated. This depends on how the final operating conditions are optimized, since the total time for elution in Figure 13D could easily have been reduced by changing the dimensions of the capillary, for example, but perhaps with some loss of resolution.

Previous investigations of biopolymer separations have tended to use some form of EOF-driven flow in capillaries, OTC, or gel-like media. Tsuda, in his excellent introduction and overview of electric field applications in chromatography, uses CEC to describe CGE and other gel-like media in CZE [29]. We believe that CGE is not really a form of CEC, but rather belongs to CE. That is, if the solid support in the capillary (or coated on the capillary surface) does not adsorb or partition the analytes, but instead acts as a sieving medium, the process is not truly chromatographic.

Despite its great potential, electroosmotically driven EC has undergone limited experimentation in the biosciences [29]. This is due, in part, to the rapid development and wide success of CZE and its related techniques. Since EC introduces a chromatographic component into the system; however, it is quite possible that electroosmotically driven EC techniques will some day outperform strictly electrophoretic procedures in a variety of biochemical applications [29].

6.6 Biopolymer Adsorption and Interaction

A lingering problem in CEC of biopolymers, especially for charged analytes, is the irreversible adsorption or interaction with charged surface functionalities. In using conventional HPLC packings, such as Spherisorb, to a certain degree, untagged (non-endcapped) silanol groups remain. Other HPLC packings, such as Vydac, have very few residual silanol groups for peptide mapping in HPLC, leaving the preferred, pure reversed-phase interactions. However, if one utilizes the Vydac packings in CEC, because there are so few free silanol groups present, residual EOF is very low [61–64]. Workers using packings such as Vydac must then apply external pressure (HPLC pump driven) to elute the proteins/biopolymers from such packings, within a reasonable period of time. These types of packings do prevent unwanted ionic biopolymer–packing interactions, but at the expense of greatly reduced EOF. In this sense, much of Lubman's work was really PEC, with questionable levels of EOF actually present, and actually approached CLC in scope and pressure-driven flow. However, if one utilizes other packings with residual silanol groups to generate reasonable EOF, then one is faced with generally unwanted, irreversible protein adsorption (interaction) with the packings, especially if RP is the desired mode of separation in CEC. How can this seeming contradiction and anomaly be overcome?

In the past, in order to prevent ionic adsorption or interaction in IEC modes (HPLC), one commonly accepted approach has been to introduce a basic (amine) salt (organic or inorganic) to compete with the analyte. However, the use of high-concentration ionic salts in CE or CEC leads to current buildup, heat generation, and eventually system shutdown at high voltages (e.g., 15 kV). Therefore, one must use low concentrations of ionic species to enable reasonable voltages and the derived EOF, but low concentrations are usually insufficient to prevent unwanted protein–silanol surface interactions. If one really wishes to perform true CEC, with little or no pressurized flow, then one must have reasonable EOF generation. That means having residual, free silanols or ion-exchange sites (e.g., sulfate or sulfonate groups), which cause their own problems in irreversible interactions with the proteins or other biopolymers.

A solution to the above method requirements might reside in the use of zwitterionic buffers, species such as Z-1 Methyl reagent (Waters Corporation) (see Table 1, p. 62). In the early days of CZE of proteins, before permanently coated capillaries were available from commercial vendors, analysts would add zwitterionic additives/reagents to the capillary buffer. These were used to provide ionic conducting solutions (ionic strength), but mainly to prevent unwanted capillary wall (silanol) interactions with the free proteins (irreversible interactions with charged surface functionalities). The zwitterions would interact with the capillary walls and the proteins, keeping one from the other by ionic repulsion. The peak shapes and efficiencies were more than adequate for simple CZE separations. The use of zwitterionic additives does not lead to heat buildup, current generation, or system shutdown in CZE. It also allows for application of high voltages and concurrent EOF, which are useful for protein analyses in CE. Thus, such approaches may well suffice in isocratic, step-gradient, or true gradient CEC in the future, without the need for pressurized flow or PEC variants.

6.7 Conclusions

It is clear that CEC and PEC are both directly applicable to a wide variety of biopolymers and that, at times, significant improvements in peak shape, plate count, resolution, efficiency, and time of analysis can be realized. The optimization of these separations, however, has yet to be fully realized. At times, workers have used pressurized flow together with EOF in PEC in order to realize reasonable analysis times, but at the expense of additional band broadening and loss of plate counts and efficiency. The wide variety of packing materials available, electric field strength, direction of voltage, voltage gradient, pH, applied external pressure, eluent composition, and overall mobile phase gradient are among the parameters that need to be optimized. CEC/PEC applications for biopolymers have yet to utilize any software chemometric approaches. For biopolymer applications—as for all CEC/PEC separations—there also remains a need for research-oriented column choices from commercial vendors to replace the nonoptimized, often improvised, packings currently in use.

The often contradicting requirements of reasonable EOF and no silanol–analyte interactions (band broadening causes) do not appear to have been resolved. The approach of introducing pressurized flow, does not solve the problem, but just forces the analytes to elute more quickly. Moreover, if there is no residual EOF, then the method is CLC. Electro-HPLC has no EOF, PEC has some residual EOF combined with pressurized driven flows and EPF, and CEC has no pressurized component to the flow. Since biopolymer classes are usually charged, applications of electro-HPLC, PEC, and CEC should find more widespread applications and positive results in the immediate future.

References

1. M. T. W. Hearn, F. E. Regnier, and C. T. Wehr (Eds.), *High-Performance Liquid Chromatography of Proteins and Peptides*, Proceedings of the First International Symposium, Academic Press, New York, 1983.
2. W. S. Hancock and J. T. Sparrow, *HPLC Analysis of Biological Compounds: A Laboratory Guide,* Marcel Dekker, New York, 1984.
3. C. Horvath (Editor), *High Performance Liquid Chromatography: Advances and Perspectives*, Volumes 1–5, Academic Press, New York, 1981–1988.
4. C. Horvath and J. G. Nikelly (Editors), *Analytical Biotechnology, Capillary Electrophoresis and Chromatography*, ACS Symposium Series 434, American Chemical Society, Washington, DC, 1990.
5. W. S. Hancock (Editor), *High Performance Liquid Chromatography in Biotechnology*, J. Wiley & Sons, New York, 1990.
6. P. R. Brown (Editor), *HPLC in Nucleic Acid Research: Methods and Applications,* Marcel Dekker, New York, 1984.

7. A. M. Krstulovic (Editor), *Nucleic Acids and Related Compounds (Parts A and B)*, CRC Press, Inc., Boca Raton, FL, 1987.

8. O. Mikes, *High Performance Liquid Chromatography of Biopolymers and Biooligomers. Part A, Materials and Techniques, Part B, Applications*, Elsevier, Amsterdam, 1988.

9. C. T. Mant and R. S. Hodges (Editors), *High-Performance Liquid Chromatography of Peptides and Proteins: Separation, Analysis, and Conformation*, CRC Press, Boca Raton, FL, 1991.

10. R. L. Cunico, K. M. Gooding, and T. Wehr. *Basic HPLC and CE of Biomolecules.* Bay Bioanalytical Laboratory, Richmond, CA, 1998.

11. C. Horvath and J. G. Nikelly (Editors), *Analytical Biotechnology: Capillary Electrophoresis and Chromatography*, ACS Symposium Series, Volume 434, American Chemical Society, Washington, D. C., 1990.

12. S. F. Y. Li , *Capillary Electrophoresis: Principles , Practice, and Applications*, Elsevier Science Publishers, Amsterdam, 1992.

13. P. D. Grossman and J. C. Colburn (Editors), *Capillary Electrophoresis- Theory and Practice*, Academic Press, Inc., San Diego, CA, 1992.

14. N. Guzman (Editor), *Capillary Electrophoresis Technology*, Marcel Dekker, New York, 1993.

15. R. Weinberger, *Practical Capillary Electrophoresis*, Academic Press, San Diego, CA, 1993.

16. J. P. Landers (Editor),*CRC Handbook of Capillary Electrophoresis: Principles, Methods, and Applications*, Second Edition, CRC Press, Boca Raton, FL, 1997.

17. D. N. Heiger, *High Performance Capillary Electrophoresis: An Introduction, A Primer, Second Edition*, Hewlett-Packard Corporation, Waldbronn, Germany, 1992.

18. D. R. Baker, *Capillary Electrophoresis*, Techniques in Analytical Chemistry Series, J. Wiley & Sons, Inc., New York, 1995.

19. P. G. Righetti (Editor), *Capillary Electrophoresis in Analytical Biotechnology*, CRC Series in Analytical Biotechnology, CRC Press, Inc., Boca Raton, FL, 1996.

20. P. Camilleri (Editor), *Capillary Electrophoresis, Theory and Practice*, CRC Press, Boca Raton, 1993.

21. R. A. Mosher and W. Thormann, *The Dynamics of Electrophoresis*, VCH Publishers, Weinheim, Germany, 1992, Chapter 7.

22. K. D. Altria and M. M. Rogan, *Introduction to Quantitative Applications of Capillary Electrophoresis in Pharmaceutical Analysis, A Primer*, Beckman Instruments, Inc., Fullerton, CA, 1995.

23. R. Weinberger and R. Lombardi, *Method Development, Optimization, and Troubleshooting for High Performance Capillary Electrophoresis*, Simon & Schuster Custom Publishing, Needham Heights, MA, 1997.

24. B. L. Karger and W. Hancock (Editors), *High Resolution Separation and Analysis of Biological Macromolecules*, Methods in Enzymology Series, Volume 270, Part A, Fundamentals, Academic Press, 1996.

25. K. D. Altria (Editor), *Capillary Electrophoresis Guidebook, Principles, Operation, and Applications*, Methods in Molecular Biology 52, Humana Press, Totowa, NJ, 1996.

26. Capillary Electrochromatography 1-day symposium, San Francisco, CA, August, 1997, organized by the California Separations Science Society, San Francisco, CA.

27. Royal Society of Chemistry, Analytical Division, North East Region, Chromatography and Electrophoresis Group, 1-day symposium on New Developments and Applications in Electrochromatography, December 3, 1997, University of Bradford, Bradford, UK.

28. R. Stevenson, K. Mistry, and I. S. Krull, *American Laboratory, 16A* August, 1998.

29. T. Tsuda (Editor), *Electric Field Applications in Chromatography, Industrial and Chemical Processes*, VCH Publishers, Weinheim, Germany, 1995.

30. M. M. Dittmann, K. Wienand, F. Bek, and G. P. Rozing, *LC/GC Magazine, 13* (10), 800 (1995).

31. J. H. Miyawa and M. S. Alesandro, *LC/GC Magazine, 16* (1), 36 (1998).

32. R. E. Majors, *LC/GC Magazine, 16* (1), 12 (1998).

33. I. H. Grant, Capillary electrochromatography. In *Capillary Electrophoresis Handbook*, Edited by K. D. Altria, Methods in Molecular Biology, Volume 52, Humana Press, Inc., Totowa, NJ, 1996, Chapter 15.

34. M. R. Euerby, C. M. Johnson, and K. D. Bartle, *LC/GC International, 39* (January, 1998).

35. M. G. Cikalo, K. D. Bartle, M. M. Robson, P. Myers, and M. R. Euerby, *Analyst, 123,* 87R, (1998).

36. W. Wei, G. Luo, and C. Yan. *American Laboratory, 20C* (January, 1998).

37. E. C. Peters, K. Lewandowski, M. Petro, J. M. J. Frechét, and F. Svec. Paper presented at HPLC 98, St. Louis, MO, May, 1998.

38. E. C. Peters, K. Lewandowski, M. Petro, F. Svec, and J. M. J. Frechét, *Anal. Commun., 35,* 83 (1998).

39. E. C. Peters, M. Petro, F. Svec, and J. M. J. Frechét, *Analytical Chemistry, 69,* 3646 (1997).

40. E. Venema, J. C. Kraak, H. Poppe, and R. Tijssen. *Chromatographia, 48* (5/6), 347 (1998).

41. A. Palm and M. V. Novotny, *Analytical Chemistry, 69,* 4499 (1997).

42. J. C. Bigge, T. P. Patel, J. A. Bruce, P. N. Goulding, S. M. Charles, and R. B. Parekh, *Analytical Biochemistry, 230,* 229 (1995).

43. C. Yang and Z. El Rassi, *Electrophoresis, 19,* 2061 (1998).

44. M. Zhang and Z. El Rassi, *Electrophoresis, 19,* 2068 (1998).

45. C. Niemann, W. Saenger, R. Nuck, and B. Pfannemuller, *Carbohydrate Res., 215,* 15 (1991).

46. Z. El Rassi, D. Tedford, J. An, and A. Mort, *Carbohydrate Res., 215,* 25 (1991).

47. J. R. Mazzeo and I. S. Krull, In *Capillary Electrophoresis Technology,* Edited by N. Guzman, Marcel Dekker, New York, 1993, Chapter 29.

48. J. R. Mazzeo, J. Martineau, and I. S. Krull, In *CRC Handbook of Capillary Electrophoresis: Principles, Methods, and Applications,* Edited by J. P. Landers, CRC Press, Inc., Boca Raton, FL, 1994, Chapter 18.

49. X. Liu, Z. Sosic, and I. S. Krull, *Journal of Chromatogrophy B, Biomedical Applications, 735,* 165 (1996).

50. I. S. Krull, J. Dai, C. Gendreau, and G. Li, *Journal of Pharm. Biomed. Analysis, 16,* 377 (1997).

51. J. M. Lin, T. Nakagama, K. Uchiyama, and T. Hobo, *Journal of Pharm. Biomed. Analysis, 15,* 1351 (1997).

52. J. M. Lin, T. Nakagama, K. Uchimaya, and T. Hobo, *Biomedical Chromatography, 11,* 298 (1997).

53. S. Li and D. K. Lloyd, *Journal of Chromatography A, 666,* 321 (1994).

54. J. M. Lin, T. Nakagama, K. Uchiyama, and T. Hobo, *Journal of Liquid Chromatography Rel. Technol., 20* (10), 1489 (1997).

55. M. R. Euerby, D. Gilligan, C. M. Johnson, S. C. P. Roulin, P. Myers, and K. D. Bartle, *Journal of Micro. Separations, 9,* 373 (1997).

56. J. M. Lin, K. Uchiyama, and T. Hobo, *Chromatographia, 47* (11/12), 625 (1998).

57. R. J. Bopp, T. J. Wozniak, S. L. Anliker, and J. Palmer, In *Pharmaceutical and Biomedical Applications of Liquid Chromatography*, Edited by C. M. Riley, W. J. Lough, and I. W. Wainer, Progress in Pharmaceutical and Biomedical Analysis, Volume 1, Pergamon Press, Elsevier Science, Ltd., Amsterdam, Holland, 1994, Chapter 10.

58. J. W. Dolan and L. R. Snyder, *American Laboratory, 50* (May, 1990).

59. L. R. Snyder, J. J. Kirkland, and J. L. Glajch, *Practical HPLC Method Development, Second Edition*, J. Wiley & Sons, New York, 1997, Chapter 10.

60. K. Schmeer, B. Behnke, and E. Bayer, *Analytical Chemistry, 67,* 3656 (1995).

61. J.-T. Wu, P. Huang, M. X. Li, M. G. Qian, and D. M. Lubman, *Analytical Chemistry, 69,* 320 (1997).

62. J.-T. Wu, M. G. Qian, M. X. Li, K. Zheng, P. Huang, and D. M. Lubman, *Journal of Chromatography A, 794,* 377 (1998).

63. P. Huang, J.-T. Wu, and D. M. Lubman, *Analytical Chemistry, 70,* 3003 (1998).

64. J.-T. Wu, P. Huang, M. X. Li, and D. M. Lubman, *Analytical Chemistry, 69,* 2908 (1997).

65. J. J. Pesek, M. T. Matyska, J. E. Sandoval, and E. J. Williamsen, *Journal of Liquid Chromatography & Related Technologies, 19* (17/18), 2843 (1996).

66. J. J. Pesek, M. T. Matyska, and L. Mauskar, *Journal of Chromatography A, 763,* 307 (1997).

67. J. J. Pesek and M. T. Matyska, *Journal of Chromatography A, 736,* 255 (1996).

68. J. J. Pesek and M. T. Matyska, *Journal of Chromatography A, 736,* 313 (1996).

69. J. Ding and P. Vouros, *Analytical Chemistry, 69,* 379 (1997).

70. J. Ding and P. Vouros, *Analytical Chemistry, 69,* 379 (1997).

71. W. A. Apruzzese and P. Vouros, *Journal of Chromatography A, 794,* 97 (1998).

72. J. Ding and P. Vouros, *American Laboratory, 16* (June, 1998).

73. B. Behnke and E. Bayer, *Journal of Chromatography A, 680,* 93 (1994).

74. *Mass Spectrometry in the Biological Sciences: A Tutorial,* Edited by M. L. Gross, Proceedings of the NATO Advanced Study Institute on Mass Spectrometry in the Molecular Sciences, Cetraro, Italy, 1990.

75. *Protein and Peptide Analysis by Mass Spectrometry,* Edited by J. R. Chapman. Humana Press, Inc., Totowa, NJ, 1996.

76. R. J. Cotter, *Time-of-Flight Mass Spectrometry, Instrumentation and Applications in Biological Research,* ACS Professional Reference Books, American Chemical Society, Washington, DC, 1997.

77. *High Resolution Separations and Analysis of Biological Macromolecules, Part A, Fundamentals, Methods in Enzymology,* Edited by B. L. Karger and W. S. Hancock, Volume 270, Section III, Mass Spectrometry.

Further Reading

Peptides and Proteins, Sugars, and Polysaccharides

G. Choudhary and C. Horvath. *Journal of Chromatography A, 781,* 161 (1997).

S. Suzuki, M. Yamamoto, Y. Kuwahara, K. Makiura, and S. Honda, *Electrophoresis, 19,* 2682 (1998).

S. K. Basak and M. R. Ladisch, *AIChE Journal, 41,* 2499 (1995).

Polymers (Synthetic Organic)

E. C. Peters, M. Petro, F. Svec, and J. M. J. Frechét, *Analytical Chemistry, 70,* 2296 (1998).

7 Method Transfer from HPLC

This chapter discusses how to transfer an existing HPLC method to a CEC or PEC format, and how the conditions may differ in CEC/PEC. Because CEC/PEC is a relatively new analytical separation technique, many analysts may wish to utilize what is described for their analyte in HPLC modes and transfer those conditions to CEC/PEC. Chapter 8 complements this discussion with basic methods development and optimization.

7.1 Differences Between CEC and PEC in Method Transfer from HPLC

PEC and CEC conditions are sufficiently different from each other that we will approach method transfer from HPLC to CEC, then to the more complicated PEC. PEC can be considered CEC but with a certain amount of pressure-driven mobile-phase movement through the packed bed. PEC is done to improve overall flow rates, especially when EOF at the applied voltage in CEC is inadequate in providing a reasonable overall analysis time [1–5]. PEC often combines pressure-driven flow with the already existing EOF. PEC is especially useful and practical when a particular HPLC packing material cannot provide sufficient EOF to permit a reasonable analysis time [6–7].

Method transfer from HPLC to pressure-driven CEC affects analysis time, plate number, and selectivity. Pressure-driven flow assistance results in worsened zone broadening in CEC but does not change the basic HPLC separation mechanism, apart from dramatically affecting the electrophoretic mobility velocity component of charged solutes.

7.2 Comparison of Basic Requirements in CEC and HPLC

CEC is nothing but HPLC with an EOF-driven mobile phase, except that the applied voltage (anode to cathode) can also influence the electrophoretic velocity [45]. Figure 1 illustrates the schematics of CEC separation using various pH values and stationary phases. Figure 1a shows a CEC separation of acids at high pH on a standard RP (C18) phase, where the analytes form anions and move by EPF towards the anode, but the overall EOF drives them eventually past the detector towards the cathode. Figure 1a also shows a packed capillary CEC arrangement, with the same EOF direction generated by the capillary wall and packing, assuming that there are anionic sites on the packing material at the operating high pH (separation of charged, anionic acids). Figure 1b shows a CEC separation of the same acids at low pH (neutral acids, uncharged now), again on a standard RP phase, where the overall EOF is reduced over Figure 1a. In Figure 1c, there is a separation of acids at low pH on a mixed-mode phase, but with a higher EOF than in Figure 1b and equal to that in Figure 1a. Thus, EOF direction and intensity are dependent not only on pH, but also on the nature of the stationary phase.

This is the primary distinction between CEC (or PEC) and HPLC; otherwise, the basic operational requirements and conditions can be identical, especially for neutral, uncharged analytes. Indeed, one can imagine the same HPLC conditions being transferred to CEC opera-

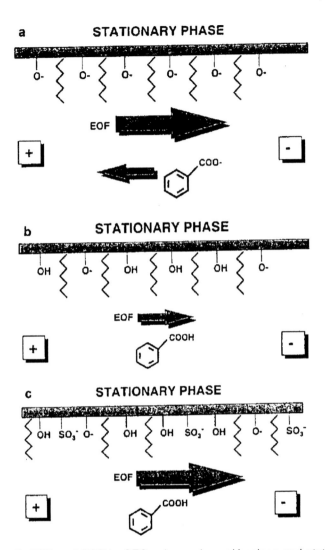

Figure 1. EPF and EOF in CEC using various pH values and stationary phases, showing possible direction and intensity of each flow for uncharged and charged analytes as a function of pH. (a) CEC separation of acids at high pH on a standard phase (RP); (b) CEC separation of acids at low pH on standard phase (RP); and (c) CEC separation of acids at low pH on a mixed-mode phase (RP and CIEC). (From Ref. [45], with permission).

tions for neutral analytes, except for the packings and the inherent differences in EOF-generating capabilities. One must still consider the important issues of conductance of the mobile phase, aqueous vs. organic composition, and concentrations of ionic buffers used to assist in the application of external voltage and generation of EOF. However, it appears that the only major factor is the nature of the stationary and the mobile phases, with voltage choice perhaps secondary.

7.3 Packings and Stationary Phases in CEC

Which packings from HPLC are suitable and desirable in CEC, and why? In CEC, as opposed to HPLC, the packing plays at least two roles: interaction with the analytes to affect resolution of the sample components and generation of EOF with the assistance of the applied voltage. The interaction of the packing with the analytes in CEC is basically the same as in HPLC; the difference has to do with EOF generation, something unimportant in HPLC. A packing that does not have free silanols or that bears negative charges, as a function of the pH of the mobile phase/buffer, will generate only insignificant EOF, lead to very long and inefficient analysis times, and be of no practical utility or use.

The CEC literature abounds with separations that have used existing HPLC packings to run CEC applications. These approaches can be considered less than ideal because the HPLC packings were never designed for this purpose. This perhaps is why several researchers report adding a pressure-driven component to their CEC applications to get enough bulk fluid flow through the capillary to elute the analytes within a reasonable time [78]. Packing materials for CEC need to be chosen not only for desirable interactions with the analyte, as in HPLC, but also to provide EOF under an applied voltage.

There is an apparent contradiction in the requirements for successful RP packings for CEC. On the one hand, a constant EOF is desired at all pH values. Unfortunately, HPCE studies have shown that EOF is a function of pH [89]. For most RP packings, then, at low pH the EOF will be minimal, while at high pH it will be maximal. There are some coated capillaries (e.g., C_{18} or C_8) used in HPCE that have been shown to have a relatively constant EOF at all pH values. However, their EOF is usually much less than for bare-faced silica capillaries. Most capillaries in HPCE and packings in CEC are silica based. The addition of sulfonic acid groups or a mixed RP/IEC packing in CEC has been shown to yield a more constant, higher EOF, mainly because of the sulfonic acid groups. However, there are some RP packings, such as the Spherisorb brand (PhaseSep, Ltd., Division of Waters Corporation) that appear to have a fair amount of untagged silanol groups, leading to an above-average EOF at other than very low pH (pH > 2).

Another problem in using uncapped silica-based packings in CEC is that these types of phases tend to generate a large number of free, anionic siloxide (from silanol) sites, which can then function as cation exchange sites for amino-containing analytes. This can lead to less than ideal peak shape (tailing), performance, and plate counts, just as in RP-HPLC. Manufacturers of HPLC packings have now gone to completely endcapped RP packings for HPLC to minimize unwanted silanolamino (analyte) interactions. These can cause poorer peak shapes in HPLC of basic analytes. It is also likely that nonionized free silanols interact with amino groups in the analytes, via hydrogen bonding, to cause unwanted peak distortions and reduced plate counts, efficiency, and resolution in both HPLC and CEC modes. If free silanol groups are needed to generate EOF but are unwanted because of their interactions with amino containing analytes, there is a contradiction in the requirements of CEC applications for such samples.

There may be ways to overcome these native problems, such as by using mixed-mode packings (cation exchange with fully endcapped RP), or using a packing that already contains an amino function as a part of the RP chain. Mixed-mode packings may still have some residual silanol groups, again to generate EOF in CEC, but with amino or amido groups tethered to the packing itselfactually a part of the packing organicsthere is less interaction with external amino-containing analytes from the sample (e.g., symmetry shield-type packings from Waters Corporation [46]. This may be one solution to the problem of free silanol groups in RP packings in CEC when analyzing for amino-containing analytes. Using just CIEC (cation ion exchange chromatography) modes in CEC, without any RP packing present, and using an acidic pH might also overcome the problems associated with amino-containing analytes, but only at the expense of selectivity and resolution. That is why for amino-containing drugs in HPLC, RP methods have become so popular compared with CIEC.

Additionally, it must be ascertained whether or not commercially packed capillaries for RP-CEC were really designed for such applications using the above criteria or were simply taken from the HPLC product line and converted. Not all HPLC packings are created equal for CEC applications, and only now are manufacturers designing and producing packings for CEC with the specific attributes suggested above. CEC in the RP mode remains in need of packings designed with the properties described above. CEC packings should generate sufficiently high EOF at the pHs needed for buffering at the applied voltage for CEC. They should also not cause undue tailing of amino-containing analytes, but should lead to good peak symmetry and plate count values, just as they should for HPLC modes.

7.4 Mobile Phases and Buffers in CEC

Methods transfer from HPLC to CEC also involves the choice of mobile phase or buffers. In CEC, the mobile phase is usually also a buffer in that it contains certain inorganic or organic salts at known concentrations in order to provide a conducting solution for generation of the EOF. The mobile phase is also crucial in terms of its pH and ability to generate free siloxide anions, which lead to EOF generation together with the applied voltage. In CEC, the mobile phase performs the same role as in HPLC in terms of partitioning of the analytes between the stationary and the mobile phases, so the requirements concerning analyte solubility, pH, organic components, and temperature are comparable. As the pure aqueous mobile phase is modified by changes in pH or the addition of organics, salts, and other additives, EOF changes as well as overall analyte mobilities, partitioning constants, and resolutions. In CEC, in contrast to HPLC, as the mobile phase is optimized, overall mobility (EPF + EOF), as well as analyte selectivity and resolutions, are changed.

In CEC, the mobile phase needs to be optimized in terms of analyte/sample solubility, generation of maximum EOF (depending on how important time is considered), selectivity, conductance, and compatibility with the stationary phase and samples injected. Transfer of an HPLC method to CEC is not as simple as transferring a given mobile phase. Nonaqueous CEC, as for nonaqueous CE, can be used, but polar solvents will be present to some extent [10–11]. It is very difficult to transfer normal-phase HPLC conditions to CEC, and much easier to transfer RP and IEC methods.

In transferring an HPLC method to CEC, one needs to retain the wanted analyte selectivity factors and provide the needed conductance properties, pH properties, ionic strength, and buffering capacity, as for CE. By adding those features necessary for CEC, the chromatographic/selectivity features from HPLC are not negatively impacted. This requires optimization of the CEC conditions; direct transfer of the exact same HPLC conditions may not be feasible. As with typical HPLC optimization methods, there is the need to perform further CEC conditions optimization with regard to pH, ionic additives, and organic composition. Chemometric software for performing CEC optimization operations is not yet commercially available, but it has been described for CE and HPLC [12–17]. The LC Resources DryLab (LC Resources, Inc., Walnut Creek, CA) software for HPLC optimization could, in theory, be applied for CEC, except for the omission of voltage optimization [18]. The literature is still developing on the subject of chemometrics for optimization of CEC conditions.

7.5 Sample Preparation Requirements

As in HPLC, there are certain important requirements that the sample must meet at the time of introduction into the CEC capillary. These requirements are very similar to HPLC sample requirements [18, 19]. The sample must be in a filtered, homogeneous, protein-free state, with current CEC packings, and in a solvent compatible with the mobile phase strength and eluting power. If preconcentration of the analyte at the head of the CEC capillary is desired, a lower-strength eluting solvent in the sample preparation can help to preconcentrate before application of the eluting solvent (isocratic/gradient). The solvent for sample preparation must be fully miscible and compatible with the mobile phase already in the capillary and that used for the separation/elution. The solvent in the sample should also be conducting, having some ionic nature (salts), a pH similar to the mobile phase, and be as much like the mobile phase as possible.

Analyte preconcentration will change some of these sample solvent parameters, but not by much. The sample should be introduced into the capillary in as small a volume as possible, in the shortest time possible, with the least amount of dispersion and band spreading possible, and in a manner that is truly representative of the sample components and their levels. The methods of sample introduction in CEC are, in general, identical to those for CE (EOF, EPF, or pressure in PEC or OT-CEC). Sample introduction by EPF alone is *not* ideal, since discrimination of the levels of analytes (especially negatively charged) entering the capillary can result. It is possible to run CEC without any EOF as long as there is an applied voltage that causes some of the analytes to migrate toward the detector (cathode) by EPF and there are differences in charge/mass ratios between such analytes. However, most CEC is not done in an EOF-free mode, since the time required for final detection could be very long.

7.6 Isocratic, Step-Gradient, and True Gradient
      -Conditions

Many HPLC conditions today utilize some type of gradient [18, 20–22].
Gradients are often utilized to reduce the total analysis time and improve resolutions, peak capacity, and peak shape. At times, isocratic,
fixed-wavelength HPLC conditions prove very useful for simpler
analyses of single or dual component samples, such as for quality
control in pharmaceutical analysis. However, because there is little
commercial instrumentation available today that utilizes gradient formation in CEC, most researchers have utilized isocratic or, at best,
step-gradient conditions [23–25]. There are some commercial CEC
instruments that can perform step-gradient or semi-gradient elutions
(Unimicro Technologies, Inc., [Pleasanton, CA]. Hewlett-Packard
[Wilmington, DE, and Palo Alto, CA], Bio-Rad Hercules (Hercules,
CA), Dionex [Sunnyvale, CA], Micro-Tech Scientific [Sunnyvale,
CA], and others).

The difficulty of transferring current gradient-elution HPLC
conditions to isocratic CEC is not easily resolved. One obvious solution would be to construct a gradient-elution CEC apparatus, and there
is ample literature that describes how this can be done [26–30]. (CEC
instrumentation is described in Chapter 4.) In such a manner, it is
possible to replicate HPLC gradient conditions in CEC, albeit with a
homemade apparatus. The Bayer approach is not true gradient CEC,
but rather pressurized electro-HPLC, although it could theoretically
perform in a true gradient CEC mode [30]. There are several other
gradient approaches described, especially by Taylor, Horvath, Myers,
Euerby, Bartle, Dorsey, and others [31–36; see also Sections 4.1.2 and
4.1.3, pp. 70–77]. An alternative approach with commercial instruments would be to transfer the gradient HPLC to a step-gradient format in CEC [23–25]. However, this transfer requires working out
entirely different step-gradient CEC conditions, such as initial/final

mobile phase compositions, time for step changes, and intermediate mobile phases. The problems with gradient formation are why most CEC applications to date have involved simple isocratic conditions, although more researchers are now moving to on-line gradient formation. There may not be a very simple method to transfer a gradient HPLC system to an isocratic CEC format and still retain separation of all interested analytes within a similar time frame. Again, the best solution would be a commercial instrument that allows isocratic, step-gradient, and gradient (linear, concave, and convex) elutions for all of CEC.

7.7 Detector Conditions

Another problem with CEC, as in most of CE today, is the much higher limits of detection (LODs) for most analytes when compared with HPLC [37]. This is due mainly to the very short path length in UV/FL, namely the capillary diameters [1216]. There are some solutions in CE and these apply in CEC, such as using a Z-cell configuration (Hewlett-Packard and LC Packings), bubble cell (Hewlett-Packard), multi-path-length cell (Hewlett-Packard), and/or preconcentration at the capillary entrance or sample stacking, as in CE [38–42].

However, in general, LODs realizable in HPLC will not be found in CEC, without additional sample preparation, selective analyte extraction and preconcentration, preconcentration at the capillary entrance (head of the packing), or modification of cell design such as has been described for CE. Quantitation at low levels of analytes in certain samples may be more problematic in CEC, as it has been for quite some time in CE, but there are practical methods and tricks (Table 1) that permit lowering of LODs in CEC so as to make the technique truly usable and useful for many samples commonly analyzed today by HPLC methods [43–44].

Table 1 Techniques to Lower Limits of Detection in CEC

Use of Increased Path Length in Detection Areas to Increase UV Absorbance

   Bubble cell with larger internal diameter

   Z-cell configuration

   Rectangular capillary dimensions

   Multi-pathlength cells with internal reflectance surfaces

Preconcentration at Head of Capillary Packing

   Analyte adsorption at head of capillary packing due to differences in sample solvent vs. running buffer or mobile phase

   Specific analyte adsorption due to affinity packing at head of packed bed[a]

   Sample stacking due to differences in ionic strength of sample solvent vs. running buffer[a]

   Selective analyte preconcentration due to hydrophobic or ionic nature of packing and analytes

   Focusing of individual analytes by capillary isotachophoresis (CITP) due to differences in isotachophoretic mobilities[b]

[a]Demonstrated in CE.
[b]Demonstrated in CITP-CZE but not yet in CITP-CEC.

7.8 Voltage Conditions

The applied voltage serves at least two important functions in CEC: first, to generate EOF and second, to produce an EPF in either direction (anode to cathode or cathode to anode) for suitably charged analyte ions. Voltage can, of course, be utilized in either CEC, where it is the sole source of bulk fluid flow, or PEC, where it is a contributor. Setting the voltage is done by experimentation in order to realize a useful and practical total analysis time and migration rate, as well as to effect an improved resolution for suitably charged species due to alterations in

7.8 Voltage Conditions

The applied voltage serves at least two important functions in CEC: first, to generate EOF and second, to produce an EPF in either direction (anode to cathode or cathode to anode) for suitably charged analyte ions. Voltage can, of course, be utilized in either CEC, where it is the sole source of bulk fluid flow, or PEC, where it is a contributor. Setting the voltage is done by experimentation in order to realize a useful and practical total analysis time and migration rate, as well as to effect an improved resolution for suitably charged species due to alterations in their EPF. The effects of voltage are directly related to pH of the mobile phase, both in terms of EOF and EPF, and these two variables are interdependent (Figure 1). EOF and EPF should really be varied as close to each other (in time) as possible, since changes in pH determine how a particular voltage setting is effective.

Ionic strength is yet another important variable to optimize and often limits the maximum voltage applicable. The nature of the ionic additives and their concentrations must be carefully regulated so that they do not cause excessive Joule heating and thereby limit the maximum voltage setting possible.

It is not always true that the highest voltage setting is the most advantageous. Higher voltages can cause increases in current, which then causes an increase in Joule heating effects, band dispersion, loss of resolution, and loss of efficiency. Lower applied voltages are more desirable if outgassing of the packed capillary at higher settings is a problem. An Ohms law plot should be obtained at the start, measuring how changes in applied voltage at a given pH will affect the current generated. In general, lower currents tend to provide improved peak shapes and resolutions, but at the expense of increased analysis times. However, higher voltages can improve resolution in terms of EPF differences between suitably charged analytes, so there can be a trade-off in this regard. In general, starting with a lower voltage setting and gradually increasing it is a better optimization approach than to start at the very highest possible voltage setting and then work downwards.

Theoretically, it is also possible to utilize voltage gradients in CEC, but this is rarely reported. Pulsed voltage has been used in capillary gel electrophoresis (CGE), especially for DNA separations; however, it is not yet generally used in CEC [43–44]. In certain applications, where improved/slower elutions into MS might be desired, pulsed-gradient CEC could be of interest. They could also be of utility for separating ionic biopolymers in SEC modes in CEC, as in CGE. The demarcation between CGE and SEC/CEC is narrow, since a packed bed (gel) in CGE is very close to a packed capillary in CEC. In neither instance is there (or should there be) any adsorption/desorption of the analytes, but rather a sieving effect and selectivity based on size or molecular weight. However, the order of elution is usually reversed: smaller before larger (CGE) or larger before smaller (SEC/CEC). CGE is generally considered to be a part of CE rather than CEC [43–44].

Besides the similarities in injection techniques for CE and CEC, automation of operations is the same for both methods since most commercial instruments are based on existing CE models. Fraction collection is also possible in CEC, and that can be done as for CE, usually as a manual approach by switching the collection vials for the catholyte reservoir as each peak elutes past the detector.

References

1. T. Eimer, K. K. Unger, and J. van der Greef, *Trends in Anaytical Chemistry, 15* (9), 463 (1996).
2. S. Ludtke, T. Adam, and K. K. Unger, *Journal of Chromatography A, 786,* 229 (1997).
3. T. Eimer, K. K. Unger, and T. Tsuda, *Fres. Journal of Analytical Chemistry, 352,* 649 (1995).
4. I. H. Grant, In *Capillary Electrophoresis Handbook*, Edited by K. D. Altria, Methods in Molecular Biology, Volume 52, Humana Press, Totowa, New Jersey, 1996, Chapter 15.
5. *Electric Field Applications in Chromatography: Industrial and Chemical Processes,* Edited by T. Tsuda, VCH Publishers, Weinheim, Germany, 1995.

10. a. A. S. Lister, J. G. Dorsey, and D. E. Burton, *Journal of High Resolution Chromatography, 20,* 523 (1997); b. Wright, A. Lister, and J. G. Dorsey, *Analytical Chemistry, 69,* 3251 (1997).

11. A. Maruska and U. Pyell, *Journal of Chromatography A, 782,* 167 (1997).

12. *Capillary Electrophoresis Guidebook: Principles, Operation, and Applications,* Edited by K. D. Altria, Humana Press, Totowa, New Jersey, 1996.

13. R. Weinberger and R. Lombardi, *Method Development, Optimization and Troubleshooting for High Performance Capillary Electrophoresis,* Simon and Schuster, Custom Publishing, Needham Heights, MA, 1997.

14. R. Weinberger, *Practical Capillary Electrophoresis,* Academic Press, Boston, 1993.

15. I. H. Grant. In *Capillary Electrophoresis Handbook,* Edited by K. D. Altria, Methods in Molecular Biology, Volume 52, Humana Press, Inc., Totowa, New Jersey, 1996, Chapter 15.

16. *Electric Field Applications in Chromatography, Industrial and Chemical Processes,* Edited by T. Tsuda, VCH Publishers, Weinheim, Germany, 1995.

17. a. J. P. Foley, Paper presented at the 1997 Frederick Conference on Capillary Electrophoresis, Frederick, MD; b. A. Coddington. Paper presented at the 1996 Frederick Conference on Capillary Electrophoresis, Frederick, MD.

18. a. L. R. Snyder, J. J. Kirkland, and J. L. Glajch, *Practical HPLC Method Development, Second Edition,* J. Wiley & Sons, New York, 1997; b. LC Resources, Inc., Walnut Creek, CA. Technical literature on Dry-Lab Software for HPLC Applications (version 2.0), 1997–98; c. L. R. Snyder and J. W. Dolan, In *Advances in Chromatography, Volume 38,* Edited by P. R. Brown and E. Grushka, Marcel Dekker, New York, 1998, Chapter 4; d. J. W. Dolan and L. R. Snyder, *LC/GC Magazine, 17* (Issue 4S), 17 (April, 1999); e. L. R. Snyder, In *New Methods in Peptide Mapping for the Characterization of Proteins,* Edited by W. S. Hancock, CRC Series in Analytical Biotechnology, CRC Press, Boca Raton, FL, 1996, Chapter 2.

19. I. S. Krull, In *Sample Preparation Technology Manual*, Zymark Corporation, Hopkinton, Mass., December, 1982.
20. C. F. Poole and S. K. Poole, *Chromatography Today*, Elsevier Science, Amsterdam, 1991.
21. B. L. Karger, L. R. Snyder, and C. Horvath, *Introduction to Separation Science*, J. Wiley & Sons, New York, 1973.
22. J. C. Giddings, *Separation Science*, J. Wiley & Sons, Inc., New York, 1991.
23. J. Ding and P. Vouros, *American Laboratory, 15* (June, 1998).
24. J. Ding and P. Vouros, *Analytical Chemistry, 69*, 379 (1997).
25. J. Ding, J. Szeliga, A. Dipple, and P. Vouros, *Journal of Chromatography A, 781*, 327 (1997).
26. C. Yan, R. Dadoo, H. Zhao, R. N. Zare, and D. J. Rakestraw, *Analytical Chemistry, 67*, 2026 (1995).
27. C. Yan, R. Dadoo, R. N. Zare, D. J. Rakestraw, and D. S. Anex, *Analytical Chemistry, 68*, 2726 (1996).
28. F. Lelievre, C. Yan, R. N. Zare, and P. Gareil, *Journal of Chromatography A, 723*, 145 (1996).
29. M. T. Dulay, C. Yan, D. J. Rakestraw, and R. N. Zare, *Journal of Chromatography A*, 725, 361 (1996).
30. B. Behnke and E. Bayer, *Journal of Chromatography A*, 680, 93 (1994).
31. M. R. Taylor and P. Teale, *Journal of Chromatography A, 768*, 89 (1997).
32. M. R. Taylor, P. Teale, S. A. Westwood, and D. Perrett, *Analytical Chemistry, 69*, 2554 (1997).
33. C. G. Huber, G. Choudhary, and C. Horvath, *Analytical Chemistry, 69*, 4429 (1997).
34. M. R. Euerby, D. Gilligan, C. M. Johnson, and K. D. Bartle, *The Analyst, 122*, 1087 (1997).
35. M. Robson, M. G. Cikalo, P. Myers, M. R. Euerby, and K. D. Bartle, *Journal of Microcolumn Sepn., 9*, 357 (1997).
36. M. Euerby, D. Gilligan, C. M. Johnson, S. C. P. Roulin, P. Myers, and K. D. Bartle, *Journal of Microcolumn Separations*, 9, 373 (1997).
37. R. Stevenson, K. Mistry, and I. S. Krull, *American Laboratory, 16*A (August, 1998).

38. Hewlett-Packard Peak Technical Newsletter, 1997. A new Z-cell UV detector for HPCE.

39. High-sensitivity detection cell for HP3D capillary electrophoresis system. Technical Note, Hewlett-Packard Company, 1997 (Publication number 12–5965–5984E).

40. CE sensitivity unmatched, HP3D CE high sensitivity cell. Hewlett-Packard Company brochure, 1996 (12–5965–7743E).

41. A. Mainka and K. Bachmann, *Journal of Chromatography A, 767,* 241 (1997).

42. M. E. Swartz and M. Merion, *Journal of Chromatography A, 632,* 209 (1993).

43. *Capillary Electrophoresis in Analytical Biotechnology,* Edited by P. G. Righetti, CRC Series in Analytical Biotechnology, CRC Press, Boca Raton, FL, 1996.

44. *CRC Handbook of Capillary Electrophoresis: Principles, Methods, and Applications, Second Edition,* Edited by J. P. Landers, CRC Press, Boca Raton, FL, 1997.

45. K. D. Altria, N. W. Smith, and C. H. Turnbull, *Journal of Chromatography B, 717,* 341 (1998).

46. U. D. Neue, T. H. Walter, B. A. Alden, Z. Jiang, R. P. Fisk, J. T. Cook, K. H. Glose, J. L. Carmody, J. M. Grassi, Y.-F. Cheng, Z. Liu, R. J. Crowley, *American Laboratory, 36* (1999).

Further Reading

1. Y. Zhang, W. Shi, L. Zhang, and H. Zou, *Journal of Chromatography A, 802,* 59 (1998).

2. P. D. A. Angus, E. Victorino, K. M. Payne, C. W. Demarest, T. Catalano, and J. F. Stobaugh, *Electrophoresis, 19,* 2073 (1998).

3. J. H. Miyawa M. S. Alasandro, and C. M. Riley, *Journal of Chromatography A, 769,* 145 (1997).

4. J. Wang, D. E. Schaufelberger, and N. A. Guzman, *Journal of Chromatographic Science, 36,* 155 (1998).

5. M. M. Dittmann and G. P. Rozing, *Biomedical Chromatography, 12,* 136 (1998).

8 Method Development and Optimization

Method development and optimization in CEC is an area that has not been widely approached in the literature [1–16]. While the same basic principles of method development and optimization used for HPLC and CE also apply to CEC [17–29], additional experimental variables in CEC need to be addressed (e.g., aqueous composition, voltage, and packings). In this chapter, method development and optimization are treated separately from true method validation, which is the focus of several texts [26–32]. However, there is some overlap between these two areas, since validation is best interwoven with the method optimization process.

The following discussion defines the specific steps that an analyst should undertake in developing and optimizing a successful method in CEC or PEC [26]. Table 1 summarizes the steps involved. Because PEC is just a pressurized mobile phase flow added on top of EOFas opposed to electro-HPLC (which is really pressurized flow with a superimposed voltage/EOF) any suggested method development/optimization/ validation approach in CEC will generally apply to PEC. The only other variable in PEC is the pressurized flow rate. Electro-HPLC method development is very similar to that for standard HPLC, with the added parameter of applied voltage.

Table 1. Overview of Steps for Method Development and Optimization

---

Analytical Method Requirements
  Who determines the requirements? What are total requirements and needs? Who will finally use these methods? When, how, and with what samples?

Analyte Standard Characterization
  Where is the standard obtained? Who synthesized it, and what were the methods of synthesis, storage, characterization, and stability?

Literature Search, Background Information, and Prior Art
  Where and how should one search, and what should one search for? (See also Table 2)

Adaptability and Choice of Possible Methods
  What do literature/technical information sources provide as initial method conditions? How can these be adapted to the present needs?

Instrumentation Preparation and Initital Studies
  What instrumentation is needed? Is the instrumentation qualified? Is it performing properly? What are the initial literature-based methods to be evaluated, and how?

Sample and Method Optimization
  What optimization steps should be employed? How can they be chemometrically based? Which parameters need to be optimized first, and why? How can sample preparation be optimized?

Demonstration of Analytical Figures of Merit with Standards and System Suitability Samples
  Under optimized conditions, demonstrate typical figures of merit, such as LOD, LOQ, linearity of calibration plots, selectivity, reporducibility, accuracy, precision, robustness, and ruggedness.

Evalutaion with Actual Samples
  Apply fully optimized CEC conditions for standards or system suitability samples to real samples of unknown levels of analyte that can be confirmed using currently accepted methods or samples of known composition.

8.1 Analytical Method Requirements

More important than any other step in CEC/PEC analytical method development is the determination of the final requirements or goals of the method to be developed. [17–22, 25–26, 33–39]. Before beginning to develop or optimize the method, it is essential to define the specific goals and needs. The goals must be in writing, established by everyone involved, and agreed upon before any experimental work is planned or undertaken.

It is essential that everyone agree on the definitions of the analytical figures of merit to be utilized. These definitions should become part of the written documentation; it is essential that no assumptions are made regarding any aspect of method development and optimization.

It is important to consider the eventual use of the method, method transfers, and the regulatory requirements that need to be met. Additional requirements might include sample throughput (time, effort, money, labor), analysis time, available instrumentation (isocratic vs. gradient elution CEC), instrument limitations (pressure, voltage, solvents), and cost per analysis.

It is important to define at the start who will eventually use the method and under what conditions. Relevant questions include the following:

- What regulatory requirements must be met for federal or state regulatory agencies?
- Will an IND or NDA be submitted with this data, and if so, what statistical requirements, good laboratory practice, and standard operating procedures must be reported and met?
- What pertinent regulatory agencies should be contacted and requested in writing for current requirements and guidelines for new analytical methods for drug analysis?
- What new requirements have been specified by the regulatory agencies involved?
- What other practical operational requirements must be met, such as available instrumentation, or other laboratory resources?

8.2 Analyte Standard Characterization

It is essential to determine as much as possible about the chemical and physical properties of the analytes, such as melting point, boiling point, solubilities, thermal stability, polarity, hygroscopicity, pK_a, pK_b, and pI. The more known about the analyte, the easier it is to actually develop a usable and practical method. It is also important to know the type of functional groups present, how they can be derivatized or stabilized, and how they can be converted into a more stable and chromatographically friendly derivative, with improved chromatographic properties.

Next, determine how the solubility and spectral properties of the analyte change with pH, and whether this property can be used for improved sample preparation and cleanup as well as for improving the separationdetection aspects of the CEC method. What are the solvents for which the analyte has the highest solubility and elution properties? Is it UV, FL, or EC active, and if so, what is known about detection properties? If these properties are unknown, they should be determined experimentally by measuring the UV and visible spectra to determine extinction coefficients and molar absorptivity as well as fluorescence properties and electrochemical activities.

It must also be determined if a standard analyte (100% pure) is available, and in what quantity. Proper storage conditions (e.g., refrigerator, room temperature, desiccator, frozen, aqueous/organic solutions) and disposal requirements must also be observed. It should be noted whether the standard is commercially available or if it will need to be synthesized. Purity must also be determined or at least documented if the standard is available as a reference material (e.g., USP, NIST).

Material safety data sheets must be archived. A documentation log should be kept related to the standard, showing receipt, origin of batch or lot, purity, and structure [31]. It is always possible that there may be more than a single component to be analyzed in the sample matrix. The number of components and possible availability of standards should be considered as well as standards of degradation products, possible impurities, and synthetic precursors.

Only those methods (e.g., MS, GC, CE, HPLC) compatible with sample stability should be considered. Instead of simply assuming homogeneity and purity of standards, one must physically demonstrate purity and chemical structure before using standards for any method development studies. It is inexcusable to assume a given purity based on what is stated on the label.

The importance of documentation in method development, optimization, and validation studies cannot be overemphasized; everything, including information from the literature, must be recorded from the beginning in a laboratory notebook [30–33].

A chain of custody for the standard should be maintained. Whenever an aliquot of the standard is removed, a notation should be made in the notebook. The purpose of the experiment and the use and fate of the aliquot should be noted. All of this information should be backed up in a separate location, safely stored and protected from loss or damage.

Meticulous record keeping is paramount because the information may become the foundation of a method, patent, or regulatory submission; without this basic documentation, patents are difficult to secure or defend. A validated method may eventually be submitted as part of a filing for FDA, EPA, or NIOSH approval and review [32]. One of the first things that any inspector asks on an inspection/site visit relates to the whereabouts of all relevant laboratory notebooks and record keeping: Where is the documentation? What is its traceability? Who handled the records? Where were they stored? Who had access to the documentation? Is there other documentation not obvious from the record?

It must be determined how many components in the sample are to be analyzed in the final analysis. Can authentic standard (reference) materials be obtained for each component? Can the sample be obtained in sufficient volume or mass?

Trace analysis presents its own unique problems, especially in terms of analyte recovery, loss, stability, and final quantitative accuracy and precision of measurements. Method development at trace levels is much more difficult and time consuming than for routine, conventional assays. Because a sample in limited supply is precious, it may be impossible to perform repeated assays. Samples in limited quantities should be assayed only after it has been shown with synthetic samples that the method is optimized for the levels of analyte expected. Never sacrifice any amount of a limited real sample until synthetic samples having the same level of analyte expected have been successfully analyzed. System suitability or placebo samples (containing various levels of the analyte of interest) are usually easy to duplicate, but real samples of unknown levels, especially from animal or human clinical studies, are much more difficult to replace.

8.3 Literature Search, Background Information, and Prior Art

The literature should be searched for all types of information related to the analyte, and it should be determined if its synthesis, physical/chemical properties, or solubility have been described, or any analytical methods developed. Consult books, periodicals, and regulatory agency compendia, such as USP/NF, AOAC, and ASTM publications. Use CAS automated/computerized literature searches, if possible. There are all sorts of literature sources to search, off- and on-line, free or for a fee (e.g., Knight-Ridder Corporation, Dialog literature services), vendor data bases, CEC column manufacturers literature, instrument vendors literature, and patent literature. A good scientific reference librarian should be utilized for accessing suitable literature sources and search routines. The Internet is a valuable source of much chemical information, with such resources as www.chemfinder.com and www.aist.go jp/RIODB/SDBS. (See Table 2 for additional Internet and other resources.)

Table 2. Literature Sources for CEC Method Development and Optimization

Traditional, Library-Type Sources

Chemical Abstracts (www.acs.org); Royal Chemical Society (UK) Abstract Service (www.rcs.uk.org); U.S. Pharmacopeial Collections of Validated Pharmaceutical Assays (www.usp.org); FDA/EPA/DOE approved analytical methods (www.fda.gov/cder/analyticalmeth.htm); I. S. Krull and M. E. Swartz. Validation Viewpoint Columns, *LC/GC Magazine*, 1997–2000

Commercial Searching Resources

Uncover®, computer-based, online literature-searching resource; Dialog Literature Searching Service, Knight-Ridder Corp., Palo Alto, CA (www.dialog.com); CASurveyor (ACS), Chromatography, CODEN: CCMAEY, ISSN: 1071-8788

HPLC/CEC Vendors

CD-ROMs such as Hewlett-Packards *The Basics of Chromatography at Your Fingertips, Capillary Electrochromatography: Technology and Applications,* and *CEC Guidebook* (1999) (www.agilent.com). CD-ROM HPLC Columns and Application Methodology, v. 2.07.125, 1999, from HPLC/CEC column vendor Advanced Separation Technologies, Inc., Whippany, NJ (www.astec.com). Websites of CEC column manufacturers, such as www.unimicrotech.com, www.biaseparations.com, www.hypersil.com, www.innovatech.com

Analytical Chemistry Software Developers

CD-ROMs and downloadable chemometrics software for HPLC method development, such as that from Advanced Chemistry Development (Toronto, ON, Canada) (www.acdlabs.com; LC Resources, Inc. (Walnut Creek, CA) (www.lcresources.com), and Intralogix, Inc. (Elmhurst, IL) (www.intralogix.com)

A literature search of an analytes structure and physical chemical properties is fundamental to any search of relevant methods and conditions. There are companies (e.g., Advanced Chemistry Development [www. acdlabs.com]) that analyze the exact structure of an analyte and use HPLC theory and fundamental approaches to predict the specific HPLC conditions that should be tried first. Other companies (e.g., Intralogix Company [www.intralogix.com]) have websites that determine the appropriate conditions by searching the existing HPLC literature. This approach draws on a very large database of HPLC conditions for specific structures, matching the conditions to the structures. Dry-Lab software (LC Resources, Walnut Creek, CA) uses chemometrics to optimize HPLC conditions but does not direct the user to which conditions should be evaluated first. An approach that combines a computerized literature search with the use of chemometric software should ultimately prove most useful.

It is impossible to overemphasize the importance of good literature use prior to method development work. Cross-reference, cross-check, and use general and specialized databases as thoroughly as possible. Searching for specific literature or company references to the parent analyte can further future efforts at improved analyte separation from matrix components and selective, specific detection with quantitation.

Determine if any analytical work on the analyte under study has been done within the company, and if so, compile the results. Communicate with others who might know more about prior analytical work performed in-house, which may be unpublished. Look for documented methods already developed, and determine where the documentation resides.

Discuss the project with others in the company or anyone who might have worked with a similar compound. Networking with others who might have devoted effort to analytical method development for the selected analyte or a close analog can prove fruitful.

8.4 Adaptability and Choice of Possible Methods

First, determine which of the methods reported in the research and regulatory literature are most amenable or adaptable to the current laboratory setting and current as well as future needs. Do not be limited to CEC methods, since CEC is a somewhat limited applications area. When searching the literature for possible CEC method conditions, one should also consider HPLC methods for the analyte of interest or a close analog (as noted in Section 8.3) and with what sample types the methods have been utilized. It may turn out that there are analytical HPLC methods reported for the analyte of interest but not in the sample matrix at hand, but this is of secondary importance.

If a CEC method already exists and meets most of the expected method requirements, the method development may be complete. Even so, there may be a need for minor modifications of the methodperhaps use of a newer C-18 CEC column, different gradient conditions, or improved (more modern) instrumentation. It is possible that methods of several years ago can be modernized, improved, and brought up to date using the very latest analytical instrumentation that meets all or most of the method requirements. This is an ideal situation, because it means a minimum amount of new work, time, and money will be expended. There is no point in starting from scratch if one can start from the current state of the art to usefully apply existing methods to meet the analytical needs. Adapting, improving, or reorienting, existing literature-based reports, is preferable to reinventing the wheel.

Determine whether it is necessary to acquire suitable instrumentation in order to reproduce, modify, improve, validate, or adapt existing methods. Sample preparation, extraction, and analyte isolation are done more cost-effectively and efficiently using a fundamental HPLC or CEC approach for the analyte standard. For maximum efficiency, it is imperative to use computer-controlled instrumentation, microprocessor driven controls, and appropriate software for routine operations.

All of the data handling should be computer based, using chromatography data systems and other commercial software for routine data handling and computing including computer-based spectral comparisons, for example, Dynamax (Varlan Corp., Palo Alto, CA) or Millennium (Waters Corp., Milford, MA). The computer must be directly interfaced with the instruments detector outputs, and there should be no manual transfer of data or manual operations after sample injection for work-up and analysis. The computer should be directly interfaced with a data backup system, laser printer, and laboratory information management system for transfer of files between computers and laboratories. Sample handling should be completely computer interfaced, including sample log-in, work-up, preparation, injection, and detection; detector output(s); and calibration plots [39].

8.5 Instrumentation Preparation and Initial Studies

Once the instrumentation and initial method conditions have been decided on, it is essential to install and test the instrumentation. Testing involves USP/FDA requirements for installation qualification (IQ), operation qualification (OQ), and performance qualification (PQ) [26]. Most analytical instrument manufacturers will provide, in writing and for a fee, IQ, OQ, and PQ specifications that guarantee that they have installed the instrument properly and that it is operating according to their specifications. This is done according to set protocols of the vendor and manufacturer, using standards of known composition to demonstrate OQ.

Initial studies can begin by preparing the analyte standard in a suitable injection solution, in known concentrations and solvents. Initial experiments are designed to determine if the CEC operating conditions are compatible with the analyte and what CEC detector responses and properties can be derived. The studies are intended to get a feeling for the conditions and how these might be further improved and modified. It is not usually expected that these first conditions will be the final method. Further optimization work is usually needed, if not for the standard, then almost surely for the samples.

As with HPLC, there are no generic conditions that will apply to all analytes. Nevertheless, we may speculate as to how one might proceed. Recognizing that pH, ionic strength, and percentage organic modifier in the run buffer are the most important variables, one should select a suitable CEC packed capillary from a reputable vendor. One might start with a C_{18}, 75-mm i.d. capillary that is at least 75 cm in effective length from injection point to detection. If one is using pressure-assisted flow, the total time for a given analysis should be shorter than for pure CEC.

Next, select a buffer, such as acetic acid + sodium hydroxide or phosphoric acid + sodium hydroxide at 10 mM, and adjust the pH to 4.0, 4.5, 5.0, 6.5, 7.0, etc. Begin scouting experiments with 80% acetonitrile in the buffer and plot retention or mobilities versus percentage organic modifier at 80%, 70%, 60%, and so forth, at several pH values. Look for values where the resolution is sufficient and run times are shorter. If a region is found where percentage acetonitrile and pH provide adequate resolution and a short overall time of analysis, then investigate the effects of variations in ionic strength and applied voltage.

Of course, this approach might work only for a reversed phase-type packing and more hydrophobic, neutral species, and not for totally ionic analytes, where a mixed mode (ion-exchange and RP) or purely ion-exchange might work much better (e.g., proteins, peptides, and amino acids).

In any case, the initial results should indicate if the original conditions provide a reasonable migration velocity and elution time for the analyte standard, whether the peak is symmetric, what plate counts are possible, and other CEC or HPLC performance features such as resolution and k' [17]. One can then proceed to improve CEC conditions in terms of total time for analysis, applied voltage, pH, gradient elution conditions, column selection, and ionic strength, using commercially available method optimization software (chemometrics) [17–18, 26]. There is not, as yet, software specifically for CEC methods development, but current HPLC software is applicable to CEC methods optimization except as far as predicting and evaluating the operational parameter of voltage.

Feasibility of the method with regard to the analytical figures of merit can now be evaluated. These figures of merit will ultimately become part of the validation process, but for the moment, they are primarily goals for optimizing CEC conditions. The solvent used for sample introduction in CEC, as in HPLC, should be identical to the running buffer to avoid peak skewing, loss of peak symmetry, precipitation of analytes, and sample stacking. It is crucial to ensure that the injection solvent used to introduce the standard analyte into the CEC instrument is 100% compatible with the mobile phases employed for the actual separation. This should be demonstrated prior to sample introduction. This can be done by visually mixing a small amount of the sample solution with an equal volume of the run buffer. There should be no cloudiness, particulates, or precipitates.

It is important to start all work with an authentic, known standard rather than a complex sample matrix. If the sample is extremely close to the standard (e.g., a bulk drug), it is sufficient to start working with the actual sample. New disposables and consumables (solvents, filters, gases, etc.) should be used. Never start method development on a CEC column with a history of prior usage, no matter how clean it may appear to be. The elution of adsorbed peaks from previous samples will lead to loss of reproducibility and jeopardize results.

8.6 Sample and Method Optimization

What if the initial analytical results are less than ideal? Then begins the process of optimization. The analyst must attempt to realize improved CEC detector conditions, utilizing the analytical figures of merit as already established for the method requirements. This may require further optimization of one operational condition (e.g., voltage, pH, column dimensions, column coverage, percent loading, temperature, or gradient conditions), or it could require the optimization of many operational parameters. To minimize the amount of work involved in further optimization, chromatographic resolution, peak shape, plate count, and other analytical figures of merit should be used as a guide.

When systematically varying one parameter at a time, whether manually or by software control, keep in mind the established goals of the separation. Whenever possible, utilize multivariate-type optimization, computer-driven, software-oriented, smart interactive type software, rather than univariate manual optimization strategies. Do not assume that the next operational condition optimized will have no effect on prior optimization results, because usually it will. Ideally, one should know how optimization of a dependent variable affects previously optimized conditions and their CEC results. Paying particular attention to experimental design and consulting the literature [17–18], change one parameter at a time and bracket sets of conditions, rather than using a trial-and-error approach. Work from an organized, methodical plan and document every step in a laboratory notebook in case of dead ends.

It is essential to work from a logical, systematic plan to select which variables are to be optimized first. The sequence depends on a variables ability to maximally affect CEC results in terms of migration velocity, elution time, resolution, plate counts, and peak symmetry. Start by varying a major variable, such as pH, organic modifier, or gradient conditions, and optimize less important variables later. Change one parameter at a time, bracket sets of conditions from one end of the scale to the other, and then narrow the the conditions to determine the final settings. More recent versions of chemometric software (e.g., DryLab, version 2.0 or higher, LC Resources, Walnut Creek, CA) permit the variation of two variables at a time in HPLC, such as temperature and solvent strength, that are interactive and interdependent [17–18, 40–41].

Keep in mind possible sample introduction requirements, such as ionic strength or pH, considered unique to a particular instrument or instrumental method of analysis [17–19]. Consider what form the sample should take for final analysis, and maintain proper sample preparation and final injection matrix compatibility with the running buffer. Sample introduction requirements mean that the final sample state should be compatible with the CEC detector requirements in terms of sample solvent, analyte concentration, homogeneity, and analyte stability. It is important that the final solution injected be fully compatible with the mobile phase, stationary phase, pH, temperature, and time required for separation/detection [19]. If there is incompatibility, change the sample solvent to one that is more compatible with the CEC mobile phase.

Sample preparation steps can involve dilution, pH adjustment, addition of organic or stability modifiers, removal of interferents, extraction of analyte from possibly interfering sample components, sample cleanup, and conditioning [19–20]. These modifications can result in a reduction in the total number of components finally present for injection, together with the analyte itself, reducing the overall resolution.

Several resources deal with sample preparation requirements for HPLC, which also apply to CEC [25, 42–43].

8.7 Demonstration of Analytical Figures of Merit with Standards and System Suitability Samples

The next stage in method development is to assess the partially optimized method with regard to the original analytical requirements (Section 8.1 and Table 3). If the requirements have not been fully realized, further optimization may be warranted.

Up to this point, all work has been performed with standards. If the required analytical figures of merit cannot be met by the standard, further analysis of samples is pointless. Once the analytical figures of merit have been optimized and documented for the standard analyte, including standardization of peak integration parameters and any treatment of the data, analysis of samples can begin.

The CEC conditions that have been optimized for the analyte standard should now be applied to a simulated synthetic sample, such as a system suitability sample (SSSs) [44]. These samples should mimic real-world samples as much as possible. They must be prepared with utmost care, because they are what will demonstrate method optimization and validation for actual samples. One prepares SSSs in the laboratory or through an outside contractor that specializes in such analytical services. The SSS is similar to samples that will follow, but the level of each analyte is precisely known. The SSSs can have fewer peaks or analytes present than the real-world samples, but they should contain the same major components and at approximately the same levels. This entire area of SSSs is examined elsewhere [44].

Table 3. Chromatographic Analytical Figures of Merit that Need to Be Met for Method Development and Optimization

Term	Definition
Peak symmetry (Asy)	Lack of skew, leading or tailing; front of peak equal to rear at midpeak height; determined by a variety of equations (see, e.g., USP, ASTM)
Plate count (N)	Total number of theoretical plates in capillary from sample injection to detection window; various equations from HPLC/GC/CE
Resolution (Rs)	Separation of peak(s) of interest from each other, no overlap at baseline, integrity of individual peaks
Height equivalent theoretical plate (HETP)	Length of capillary divided by total number of theoretical plates, as in HPLC/GC/CE
Electroosmotic flow (EOF)	Rate of bulk fluid flow through CEC capillary (cm/min or vol./min)
Retention factors	Rate and time (volume) of elution for an analyte peak, as defined for HPLC/GC
Specificity[a]	Separation of analyte of interest from other peaks in sample; identification of peak by retention time, migration time, and elution time/volume
Linearity and range of calibration plot[a]	Plot of concentration vs. peak height or area and correlation with a straight line; coefficient of linearity; range of linear portion of the calibration plot
Limit of detection[a]	Smallest amount of analyte detectable above noise level at a specified ratio, such as 2:1 or 3:1 (see, e.g., USP, ICH, ASTM)
Limit of quantitation[a]	Smallest amount of analyte that can be accurately and precisely quantitated at a set signal-to-noise ratio (e.g., 10:1)
Accuracy[a]	Measure of the exactness of an analytical method; percentage of analyte recovered by assay
Precision[a]	Measure of the degree of repeatability of an analytical method under normal operations
Robustness[a]	Stability of method results or CEC performance (k′, N, Rs, Asy, HEPT

[a]Validation criterion.
From Ref. [20-22]

At the onset of sample analysis, peak homogeneity must be documented by performing appropriate peak identification studies (e.g., PDA, MS, FTIR). Peak homogeneity suggests peak purity, but it is also possible for two coeluting compounds to demonstrate peak homogeneity if they coelute exactly. Peak purity generally refers to the presence of a single compound rather than a mixture of two unresolved components. Include blank and control samples along with a well-defined SSS [26–30]. Finally, ensure the absence of false positives or false negatives arising from artifacts in the method, chemicals/solvents used, and instrumental aberrations.

8.8 Evaluation with Actual Samples

Once a CEC method has been optimized for standards and SSSs, it is time to work with actual samples. In working with actual samples to derive analytical figures of merit, it is necessary to know the exact level of analytes present; otherwise quantitation is, at best, very difficult. In addition to using real samples, it is necessary to perform analyses with blanks, controls, and placebos.

Depending on the sample, it may be necessary to develop sample-specific preparation and cleanup methods that vary from those initially developed. This may be needed for very complex samples such as biofluids, blood, and plasma. The goal is to minimize the time, cost, and effort required per sample while still satisfying the CEC detector requirements. Efficiency is enhanced and results improved if numerous sample analyses can be performed using the same CEC column and sample preparation conditions, and so that the analyte peaks are baseline resolved from all other remaining, potentially interfering peaks. This will permit identification and quantitation of the analyte peak(s), with much improved accuracy and precision for all future measurements.

The final sample should be fully compatible with the CEC detector requirements, with the analyte peak in clear, unobstructed sight, ready for quantitation, and ideally using an external standard calibration plot or standard additions. Sample compatibility is especially important for CEC, where the capillary packing is not very forgiving, does not tolerate many impurities or crudeness of the sample, and is reactivated only with great difficulty if clogged or contaminated. If the sample peak is not 100% pure and homogeneous, further method optimization and/or sample preparation may be required [19–20, 42, 43].

It is up to the analyst to evaluate experimentally which of many different quantitation methods possible works best with a particular sample [26, 44–46]. Table 4 summarizes the most commonly utilized methods of quantitation for the determination of levels of a given analyte [33–39].

Table 4. Methods of Quantitation Useful in CEC

Type	Description	Pros and cons
External standard	Generation of a calibration plot of known concentration standard vs. peak heights/areas on a detector.	Requires knowledge of % recovery. Requires approximate level of analyte in sample, linear plot for area/height.
Internal standard	Requires a well-behaved standard having similar chemical and physical properties: k', retention times, peak height/area.	Preparation of calibration plot corrects for percentage recovery and matrix effects, losses of analyte.
Isotope dilution	Requires a well-characterized isotopic standard spiked to sample before work-up or analysis.	MS detection; corrects for everything: percentage recovery, sample loss, matrix effects, errors; almost foolproof.
Standard addition	Spike at ½X, X, 2X levels; plot and extrapolate to determine original analyte level.	Corrects for matrix effect; good confirmation technique but not a validation technique.
Recovery study	Spike known levels of analyte into blank matrix.	Necessary to demonstrate percentage recovery; not a validation technique.

A selective sample preparation method results in a purer final injection solution, which in turn results in better accuracy, precision, reproducibility, and final validation with real samples. At the same time, sample preparation should be simple, perhaps a simple dilution scheme, to minimize time and cost of analysis.

References

1. I. H. Grant, In *Capillary Electrophoresis Handbook*, Edited by K. D. Altria, Methods in Molecular Biology, Volume 52, Humana Press, Inc., Totowa, NJ, 1996, Chapter 15.
2. *Electric Field Applications in Chromatography, Industrial and Chemical Processes*, Edited by T. Tsuda, VCH Publishers, Weinheim, Germany, 1995.
3. M. Robson, M. G. Cikalo, P. Myers, M. R. Euerby, and K. Bartle, CEC: A review. *Journal of Microcolumn Separations, 9*, 357 (1997).
4. M. Euerby, D. Gilligan, C. M. Johnson, S. C. P. Roulin, P. Myers, and K. D. Bartle, *Journal of Microcolumn Separations, 9*, 373 (1997).
5. M. M. Dittmann, K. Wienand, F. Bek, and G. P. Rozing, *LC/GC Magazine, 13* (10), 800 (1995).
6. M. M. Dittmann and G. P. Rozing, *Journal of Chromatography A, 744*, 63 (1996).
7. G. Ross, M. Dittmann, F. Bek, and G. Rozing, *American Laboratory, 34* (March, 1996).
8. M. M. Dittman, K. Wienand, F. Bek, and G. P. Rozing, *LC/GC Magazine, 13* (10), 800 (1995).
9. New CE instrument adds capillary electrochromatography. Hewlett-Packard Peak Technical Newsletter, January, 1996.
10. M. R. Euerby, C. M. Johnson, K. D. Bartle, P. Myers, and S. C. P. Roulin, *Anal. Commun., 33*, 403 (1996).
11. R. Weinberger, *American Laboratory, 24 F*F (May, 1997).
12. R. Stevenson, *American Laboratory, 24 H* (May, 1997).
13. J . H. Miyawa and M. S. Alesandro, *LC/GC Magazine, 16* (1), 36 (1998).
14. R. E. Majors, *LC/GC Magazine, 16* (1), 12 (1998).

15. R. Dadoo, C. Yan, R. N. Zare, D. S. Anex, D. J. Rakestraw, and G. A. Hux, *LC/GC Magazine, 15* (7), 630 (1997).
16. L. A. Colon, Y. Guo, and A. Fermier, *Analytical Chemistry, News & Features, 461A* (August 1, 1997).
17. a. L. R. Snyder, J. J. Kirkland, and J. L. Glajch, *Practical HPLC Method Development, Second Edition,* J. Wiley & Sons, New York, 1997; b. LC Resources, Inc., Walnut Creek, CA. Technical literature on DryLab Software for HPLC Applications (version 2.0), 1997–98.
18. a. L. R. Snyder and J. W. Dolan, In *Advances in Chromatography, Volume 38,* Edited by P. R. Brown and E. Grushka, Marcel Dekker, New York, 1998, Chapter 4; b. J. W. Dolan and L. R. Snyder , *LC/GC Magazine* Supplement, Volume 17 (Issue 4S), 17 (April, 1999); c. L. R. Snyder, In *New Methods in Peptide Mapping for the Characterization of Proteins,* Edited by W. S. Hancock, CRC Series in Analytical Biotechnology, CRC Press, Boca Raton, FL, 1996, Chapter 2.
19. I. S. Krul,. In *Sample Preparation Technology Manual,* Zymark Corporation, Hopkinton, Mass., December, 1982.
20. C. F. Poole and S. K. Poole, *Chromatography Today*, Elsevier Science Publishers, Amsterdam, 1991.
21. B. L. Karger, L. R. Snyder, and C. Horvath, *Introduction to Separation Science,* J. Wiley & Sons, New York, 1973.
22. J. C. Giddings, *Separation Science*, J. Wiley & Sons, Inc., New York, 1991.
23. I. S. Krull, J. R. Mazzeo, and C M. Selavka, *Biomed. Chromatography, 6,* 259 (1992).
24. P. J. Schoenmakers and M. Mulholland, *Chromatographia, 25* (8), 737 (1988).
25. B. A. Bidlingmeyer, *Practical HPLC Methodology and Applications,* Wiley-Interscience, New York, 1992.
26. M. E. Swartz and I. S. Krull, *Analytical Method Development and Validation,* Marcel Dekker, New York, 1997.
27. *Capillary Electrophoresis Guidebook, Principles, Operation, and Applications,* Edited by K. D. Altria, Humana Press, Totowa, New Jersey, 1996.
28. M. E. Swartz, *Journal of Liquid Chromatography, 14* (5), 923 (1991).

29. G. M. McLaughlin, J. A. Nolan, J. L. Lindahl, R. H. Palmieri, K. W. Anderson, S. C. Morris, J. A. Morrison, and T. J. Bronzert, *Journal of Liquid Chromatography and Related Technologies, 15* (6/7), 961 (1992).

30. M. E. Swartz and I. S. Krull, *Pharmaceutical Technology Magazine, 104* (March, 1998).

31. I. S. Krull and M. E. Swartz, *LC/GC Magazine, 15* (12), 1122 (1997).

32. I. S. Krull, M. Swartz, and C. Gendreau, *Todays Chemist at Work*, an ACS Publication, 26 (February, 1997).

33. D. A. Skoog and J. J. Leary, *Principles of Instrumental Analysis, Fourth Edition*, Saunders College Publishing, 1992, Chapter 26.

34. S. E. Manahan, *Quantitative Chemical Analysis*, Brooks/Cole Publishing Company, Monterey, CA, 1986.

35. R. de Levie, *Principles of Quantitative Chemical Analysis*, McGraw-Hill, New York, 1997.

36. A. L. Day, Jr., and A. L. Underwood, *Quantitative Analysis, Fifth Edition*, Prentice-Hall, Englewood Cliffs, NJ, 1986.

37. R. M. Smith, *Gas and Liquid Chromatography in Analytical Chemistry*, John Wiley & Sons, New York, 1988.

38. J. W. Robinson, *Undergraduate Instrumental Analysis, Fifth Edition*, Marcel Dekker, New York, 1995.

39. *Handbook for Instrumental Techniques for Analytical Chemistry*, Edited by F. Settle, Prentice Hall, PTR, Upper Saddle River, NJ, 1997.

40. I. Molnar, J. W. Dolan, and L. R. Snyder. Paper presented at HPLC 97, Birmingham, UK, June, 1997.

41. L. R. Snyder and J. W. Dolan, The Linear-Solvent-Strength Model of Gradient Elution, In *Advances in Chromatography, Volume 38*, Edited by P. R. Brown and E. Grushka, Marcel Dekker, New York, 1998, Chapter 4.

42. R. W. Frei and K. Zech (Editors), *Selective Sample Handling and Detection in High-Performance Liquid Chromatography*, Elsevier, Amsterdam, Holland, Parts A, 1988, and B, 1989.

43. H. Lingeman and W. J. M. Underberg, In H. Lingeman and W. J. M. Underberg (Editors), *Detection-Oriented Derivatization Techniques in Liquid Chromatography*, Marcel Dekker, New York, 1990.

44. I. S. Krull and M. E. Swartz, *LC/GC Magazine, 17* (3), 244 (1999).

45. a. I. S. Krull and M. E. Swartz, *LC/GC Magazine*, *16* (10), 922 (1998);
  b. I. S. Krull and M. E. Swartz, *LC/GC Magazine*, *16* (12), 1084
  (1998).

46. FDA, Guideline for Submitting Samples and Analytical Data for
  Methods Validation, US Government Printing Office, February, 1997,
  1990–281–794:20818.

Further Reading

1. Y. Zhang, W. Shi, L. Zhang, and H. Zou, *Journal of Chromatography A*, *802*, 59 (1998).

2. P. D. A. Angus, E. Victorino, K. M. Payne, C. W. Demarest, T. Catalano, and J. F. Stobaugh, *Electrophoresis*, *19*, 2073 (1998).

3. J. H. Miyawa, M. S. Alesandro, and C. M. Riley, *Journal of Chromatography A*, *769*, 145 (1997).

4. J. Wang, D. E. Schaufelberger, and N. A. Guzman, *Journal of Chromatographic Science*, *36*, 155 (1998).

5. M. M. Dittmann and G. P. Rozing, *Biomed. Chromatography*, *12*, 136 (1998).

9 | Conclusions and Future Developments

T here have been any number of analytical techniques that, at first, appeared extremely appealing and filled with promise. Supercritical fluid chromatography (SFC) could be considered as one such technique in separations science, which suffered a somewhat ignominious demise [1–2]. Even HPCE has never really reached the status of HPLC and GC, and there has been a decreasing emphasis on its commercialization in the past several years.

Capillary techniques, of course, suffer from not being preparative in nature or even semipreparative, and thus they immediately lose such applications and utilizations. And although capillary techniques offer such advantages as smaller sample requirements, smaller volumes of buffers required, ability to use exotic and expensive solvents or reagents, and better sensitivity; these advantages have not yet carried even capillary LC (CLC), available for the past 25 years, to the heights of commercial and industrial success. But there appear to be increasing interest and applications of CLC, especially for CLC-ESI-MS, and thus today this area is undergoing a renaissance. There are companies dedicated to CLC and related instrumental techniques (e.g., Waters Corporation [www.waters.com], Microtech, and LC Packings [lcpack-ings.com)].

Microbore LC became available perhaps 20 years ago also, and it too had large numbers of proponents, some commercialization efforts, and an instrumental market, but it too appears to have fallen by the wayside.

9.1 Applications as the Driving Force for CEC

What decides whether a new analytical technique will grow, prosper, and become widely accepted and applied? It must have demonstrated applications for real-world samples. The technique must be able to analyze, qualitatively and quantitatively, samples that are already ana-lyzed by existing methods, and it must do these better, faster, cheaper, or with better resolution, automation, and reduced sample preparation, and with equal or better accuracy and precision. Fortunately, there are a number of pharmaceutical laboratories, especially in the United King-dom, where applications work with CEC, as well as some fundamental work on gradient elution CEC-MS, is taking place, often with exciting results (e.g., Astra Charnwood, Glaxo Wellcome, and Zeneca) [3–7]. Much of the work thus far has dealt with pharmaceuticals, low-molecu-lar-weight analytes, at times in crude biofluid matrices, but it is not yet clear that such samples are being routinely analyzed by CEC or any variations thereof [4–5, 8–9]. The real question has to do with what types of unique samples or applications will CEC/PEC serve best. If there are not enough of these applications or samples, will this new technique really shine?

9.2 Analysis of Complex Samples by CEC

What is the value of high plate counts (N)? First, large N values lead to high peak capacity, which enables separation of more complex mix-tures. Peak capacity refers to how many peaks can be fit into a typical separation time frame for a given column length. Naturally, analysts prefer to have high peak capacities, which enable resolution of multiple peaks in a complex mixture. High peak capacities usually come about by narrowing peak volumes or peak dimensions, with improved peak symmetry values, such that each peak elutes in a minimum volume with improved signal-to-noise ratio. All of these noble attributes have already been demonstrated, in many laboratories, for CEC/PEC [45, 8, 10–11].

With HPLC, analysis of complex samples, such as amino acids and polychlorinated biphenyls, suffers from coelution of some of the components. A general lack of resolution makes it difficult to qualitatively determine the presence of a particular analyte and demonstrate peak purity. Increasing the plate count may increase the peak capacity by the square root. A three- to fivefold improvement in peak capacity is not a big factor but can extend the operating range, and hence sample complexity, from 60 to perhaps 100 components. For example, tryptic digests of proteins are designed to show the presence of new peaks, indicating that the organism has mutated or reverted. A chromatogram with three times the peak capacity would be expected to show a new peak characteristic of this event more obviously [12]. For simple samples, the number of available plates might be traded for shorter analysis times [3]. These are all very real possibilities in CEC, since it offers significantly higher peak capacities than conventional HPLC. Further, this is precisely the advantage that capillary gas chromatography (CGC) enjoyed over packed GC columns. It took 30 years, but today, capillary columns account for nearly 75% of GC separations [13].

In addition to improved plate counts and peak capacities, leading to improved resolutions for very complex samples with many components, CEC offers the added resolving uniqueness of EPF (mobility) differences for suitably charged or able-to-be-charged analytes, especially proteins/peptides. Similar to HPCE and MEKC, CEC should show the greatest advantages and attributes for charged analytes, especially where the small differences in size, molecular weight, charge, solubility, and hydrophobicity make it difficult to separate such similar components by conventional HPLC or HPCE alone.

9.3 Specialized Applications

CEC is a niche player for the time being, and although it will be able to solve certain analytical problems better than existing techniques, such applications may be far less than one would like to imagine. Since CEC has come along after the evolvement and development of so many other well-accepted separation techniques, it may have a hard time gaining acceptance and market share of applications [8–9].

Future applications might include peptide maps, mixtures of proteins, mixtures of amino acids or aminoglycosides, or complex mixtures of drug metabolites (polar, charged). CEC does not appear to be a technique ideally suited for mixtures of neutral species, such as PAHs, where GC and HPLC methods work very well and are fully satisfactory by now [14–17]. Interestingly, published reports on CEC applications rarely make a direct comparison between the currently accepted method and CEC.

While there may be applications where CEC or PEC will readily outperform other current analytical/separation methods, improved MS methods may make inroads into some of CEC's unique applications. Very complex samples or very difficult to resolve analytes in the same sample matrix may well be served by CEC and/or a combination of CEC-MS, which could provide resolutions above and beyond CEC or MS alone. However, those types of applications would not provide a large number of repeat applications that could drive the CEC market share and acceptance rate. Given that CEC has been piggybacked onto HPCE instrumentation, in the hopes of selling more such instruments, this would appear to be the consensus among major instrument manufacturers as well. CEC may thus remain basically a research tool or technique, useful for a smaller number of nonroutine applications, providing unusual resolving capabilities for just those samples needing such advantages.

9.4 The Ideal Capillary, Liquid Separations Technique

Finally, the combination of all capillary separation techniques into one single instrumental platform, permitting switching from HPCE to CLC to CEC to PEC, at will and with little downtime, is appealing [8–9, 18–19]. This type of an instrument would permit the combination of pressurized flow alone (HPLC pump for CLC), voltage applications alone (HPCE, CEC), and a combination of pressure- and EOF-driven flows (PEC, electro-HPLC). This is the ideal capillary liquid separations instrument, for it would allow an operator, with suitable computer software, to automatically vary instrumental methods, almost at will, with chemometric software routines, and compare the separations possible with each of these capillary liquid separation domains. At the same time, the availability of suitable chemometric software, such as DryLab for HPLC, CE, CEC, or CLC, would permit the analyst to automatically vary certain parameters, again almost at will, until the unique, idealized set of operational conditions is found. The future analyst would be able to optimize not only the set of operational parameters for a single separation technique but would then also be able to determine the optimal separation technique, even those capillary techniques yet to be defined.

References

1. L. R. Snyder, J. J. Kirkland, and J. L. Glajch, *Practical HPLC Method Development, Second Edition*, J. Wiley & Sons, New York, 1997.
2. C. F. Poole and S. K. Poole, *Chromatography Today*, Elsevier Science Publishers, Amsterdam, 1991.
3. J. Ding and P. Vouros, *Analytical Chemistry, 69*, 379 (1997).
4. M. Robson, M. G. Cikalo, P. Myers, M. R. Euerby, and K. Bartle, *Journal of Microcolumn Separations, 9*, 357 (1997).
5. M. Euerby, D. Gilligan, C. M. Johnson, S. C. P. Roulin, P. Myers, and K. D. Bartle, *Journal of Microcolumn Separations, 9*, 373 (1997).
6. S. J. Lane, R. Boughtflower, C. Paterson, and M. Morris, *Rapid Comm. Mass Spec., 10*, 733 (1996).

7. S. J. Lane, R. Boughtflower, C. Paterson, and T. Underwood, *Rapid Comm. Mass Spec.*, *9*, 1283 (1995).

8. R. Stevenson, K. Mistry, and I. S. Krull, *American Laboratory*, *16A* (August, 1998).

9. I. S. Krull, G. Li, and M. E. Swartz, Paper #144 presented at the November 16–21, 1997 Eastern Analytical Symposium, Garden State Convention Center, Somerset, NJ.

10. N. W. Smith and M. B. Evans, *Chromatographia*, *41* (3/4), 197 (1995).

11. N. W. Smith and M. B. Evans, *Chromatographia*, *38*, 649 (1994).

12. M. E. Szulc, R. Mhatre, J. Mazzeo, and I. S. Krull, In *High Resolution Separation of Biological Macromolecules, Methods in Enzymology Series*, Edited by B. L. Karger and W. Hancock, Academic Press, Chapter 8, p. 175 (1996).

13. J. H. Knox, Personal communication, November, 1997.

14. M. M. Dittmann, K. Wienand, F. Bek, and G. P. Rozing, *LC/GC Magazine*, *13* (10), 800 (1995).

15. M. M. Dittman and G. P. Rozing, *Journal of Chromatography A*, 744, 63 (1996).

16. G. Ross, M. Dittmann, F. Bek, and G. Rozing, *American Laboratory*, *34* (March, 1996).

17. M. M. Dittman, K. Wienand, F. Bek, and G. P. Rozing, *LC/GC Magazine*, *13* (10), 800 (1995).

18. Cs. Horvath, Presentation made at Waters Corporation, Milford, MA, USA, November, 1997.

19. A. S. Rathore and Cs. Horvath, *Journal of Chromatography A*, *743*, 231 (1996).

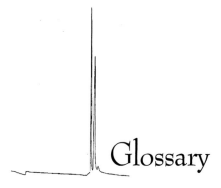# Glossary

ACN = acetonitrile
AGP = α-1-acid glycoprotein
aka = also known as
APS = (3-aminopropyl) trimethoxysilane
AMPS = 2-methyl-1-propanesulfonic acid
amu = atomic mass units
AOAC = Association of Official Analytical Chemists
ASTM = American Society for Testing and Materials

BSA = Bovine serum albumin

CAS = Chemical Abstracts Service
βCD = β-cyclodextrin

CE = capillary electrophoresis
CEC = capillary electrochromatography

CGC = capillary gas chromatography
CGE = capillary gel electrophoresis
CIT = capillary isotachophoresis
CLC = capillary liquid chromatography
CIEC = cation ion exchange chromatography
CZE = capillary zone electrophoresis

DNA = deoxyribonucleic acid

EK = electrokinetic injection
electro-HPLC = the application of applied voltage in capillary liquid chromatography
EOF = electroosmotic flow
EPA = U. S. Environmental Protection Agency

EPF = electrophoretic force
ESI = electrospray ionization
ex/em = excitation/emission wavelengths

FDA = U. S. Food and Drug Administration
FL = fluorescence detection

Glc1 = monoglucoside (one glucose monomer)
Glc3 = triglucoside (three glucose monomers in series, an oligosaccharide of three repeating units)
Glc6 = hexaglucoside (sixglucose monomers in series, an oligosaccharide of six repeating units)
GC = gas chromatography
GLP = good laboratory practice
GPC = Gel permeation chromatography

HETP = height equivalent to a theoretical plate
HPβCD = hydroxypropyl-β-cyclodextrin
HPLC = high-performance liquid chromatography
HPCE = high-performance capillary electrophoresis
HIC = hydrophobic interaction chromatography
HP = Hewlett-Packard Corp.
HPCE = high-performance capillary electrophoresis
HSA = human serum albumin

I = current
ICH = International Conference on Harmonization
ID = internal diameter
IEC = ion-exchange chromatography
IQ = installation qualification
ITS = ion trap storage MS

kV = kilovolt
k´ = capacity factor

L = total length of capillary, end to end (anode to cathode)
l = effective length of capillary from injection end to on-line detector
LIF = laser induced fluorescence (detection)
LIMS = laboratory information management system
LOD = limits of detection
LOQ = limits of quantitation (or quantification)

m = meter
MECC = micellar electrokinetic capillary chromatography (aka MEKC)
MEKC = micellar electrokinetic chromatography (aka MECC)
MES = 2-(N-morpholino)ethanesulfonic acid
MIP = molecularly imprinted polymers
mol = mole
MS = mass spectrometry or spectrometer

MW = molecular weight

N = theoretical plates or efficiency

n = number of repeat measurements

NF = National Formulary

NPS = nonporous silica

NIOSH = National Institute of Occupational Safety and Health

PAH(s) = polycyclic aromatic hydrocarbon(s)

NIST = National Institute of Standards and Technology

ODS = octadecylsilane

OQ = operational qualification

OT = open tubular

OTC = open tubular chromatography

OT-CEC = open tubular CEC

PCBs = polychlorinated biphenyls

PEC = pressurized flow CEC

PMβCO = permethyl-β-cyclodextrin

PQ = performance qualification

PTH = phenyl thiohydantoin

pI = isoelectric point

QC = quality control

QA = quality assurance

RAM = random access memory

RP = reversed phase

RPC = reversed phase chromatography

Rs = resolution

RSD = relative standard deviation

% RSD = percent RSD

SFC = supercritical fluid chromatography

SOP = standard operating procedure

SSS = system suitability sample

SRM = standard reference material

TEA = triethylamine

TEOA = triethanolamine

TES = triethoxysilane

TIC = total ion current chromatogram

TOFMS = time-of-flight mass spectrometry

USP = United States Pharmacopeia

UV = ultraviolet detection

μA = micro amps

μM = micromolar

V = volts

v/v = volume to volume ratio

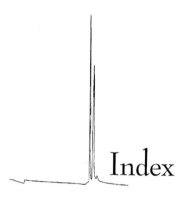

Index